Ekhlas Mohammed

Modulate cytogenetic effects of benzoic acid by pomegranate molasses

Ekhlas Mohammed

Modulate cytogenetic effects of benzoic acid by pomegranate molasses

Investigation of cytogenetic effect of pomegranate molasses and its ability to modulate genotoxic effect of benzoic acid

LAP LAMBERT Academic Publishing

Impressum / Imprint

Bibliografische Information der Deutschen Nationalbibliothek: Die Deutsche Nationalbibliothek verzeichnet diese Publikation in der Deutschen Nationalbibliografie; detaillierte bibliografische Daten sind im Internet über http://dnb.d-nb.de abrufbar.

Alle in diesem Buch genannten Marken und Produktnamen unterliegen warenzeichen-, marken- oder patentrechtlichem Schutz bzw. sind Warenzeichen oder eingetragene Warenzeichen der jeweiligen Inhaber. Die Wiedergabe von Marken, Produktnamen, Gebrauchsnamen, Handelsnamen, Warenbezeichnungen u.s.w. in diesem Werk berechtigt auch ohne besondere Kennzeichnung nicht zu der Annahme, dass solche Namen im Sinne der Warenzeichen- und Markenschutzgesetzgebung als frei zu betrachten wären und daher von jedermann benutzt werden dürften.

Bibliographic information published by the Deutsche Nationalbibliothek: The Deutsche Nationalbibliothek lists this publication in the Deutsche Nationalbibliografie; detailed bibliographic data are available in the Internet at http://dnb.d-nb.de.

Any brand names and product names mentioned in this book are subject to trademark, brand or patent protection and are trademarks or registered trademarks of their respective holders. The use of brand names, product names, common names, trade names, product descriptions etc. even without a particular marking in this work is in no way to be construed to mean that such names may be regarded as unrestricted in respect of trademark and brand protection legislation and could thus be used by anyone.

Coverbild / Cover image: www.ingimage.com

Verlag / Publisher:
LAP LAMBERT Academic Publishing
ist ein Imprint der / is a trademark of
OmniScriptum GmbH & Co. KG
Heinrich-Böcking-Str. 6-8, 66121 Saarbrücken, Deutschland / Germany
Email: info@lap-publishing.com

Herstellung: siehe letzte Seite /
Printed at: see last page
ISBN: 978-3-659-49944-9

Copyright © 2015 OmniScriptum GmbH & Co. KG
Alle Rechte vorbehalten. / All rights reserved. Saarbrücken 2015

Investigation of cytogenetic effect of pomegranate molasses (PM) and its ability to modulate the genotoxic effect of benzoic acid (E210)

by
Author
EKHLAS. Mohammed .F. Al –Tai
Master Science in College of Agriculture , University of Baghdad
Baghdad - Iraq

Contents

Article No.	Article Title	Page No.
	Introduction	1
1-1	Pomegranate	4
1-1-1	Description	4
1-1-2	Habital	4
1-1-3	Traditional Use	5
1-1-4	Chemical composition of pomegranates	5
1-2	Pomegranate fruit derived products	7
1-3	pomegranate molasses	9
1-3-1	Antioxidant activity	10
1-3-2	The health benefits of pomegranate derived products.	11
1-4	Food Additives	13
1-4-1	Classification of food additives	14
1-4-1-1	Preservatives	15
1-4-1-2	Antimicrobials	15
1-4-2	Effect of additives in food	18
1-5	Benzoic acid	18
1-5-1	Effect of Benzoic acid	19
1-6	Cytogenetic analyses	20
1-6-1	Mitotic index (MI)	21
1-6-2	Micronucleus assay (MN)	22
1-6-3	Chromosome aberrations (CA)	22

1-6-4	Replicative index assay (RI)	23
2-1	Materials and methods	25
2-1-1	Equipments and Apparatus	25
2-1-1	Chemical Material	26
2-2	Methods	27
2-2-1	Preparation process of pomegranate molasses	27
1-2-2	Benzoic acid preparation	27
1-2-3	Preparation for cytogenetic study on human blood lymphocyte	27
1-2-3-1	Antibiotic solutions	27
1-2-3-2	Growth medium	28
1-2-3-3	Colchicines solution	28
2-2-3-4	Hypotonic potassium chloride solution	28
2-2-3-5	The fixative	28
2-2-3-6	Sorenson's buffer	28
2-2-3-7	Giemsa stain	28
2-2-3-8	Phytohaemagglutinin (PHA)	29
2-2-4	The Preparation of Human Chromosome aberrations from Peripheral Blood	29
2-2-4-1	Blood collection	29
2-2-4-2	Blood culturing	29
2-2-5	Cytogenetic experiments	30
2-2-6	Slide preparation	31
2-2-7	Staining	31

2-2-8	Cytogenetic parameter Analysis scoring	31
2-2-9	Statistical analysis	32
3	Results and Discussion	33
3-1	Cytogenetic effects of pomegranate molasses (PM) extracts and benzoic acid (E210) in vitro	33
3-1-1	Effects of pomegranate molasses (PM) and benzoic acid (E210) on mitotic index:	33
3-1-2	Effects of pomegranate molasses (PM) extract and benzoic acid (E210) on chromosomal aberrations in lymphocytes culture.	34
3-1-3	Effects of pomegranate molasses (PM) extract and benzoic acid (E210) on Micronucleus (MN) formation.	35
3-1-4	Effects of pomegranate molasses (PM) extract and benzoic acid (E210) on replication index (RI)	35
3-2	Interaction between benzoic acid an pomegranate molasses extract on blood lymph	37
3-2-1	Pre –benzoic acid treatment with pomegranate molasses extract	37
3-2-2	Post-benzoic acid treatment with pomegranate molasses extract	38
3-2-3	Simultaneous treatment benzoic acid with pomegranate molasses extract	38
	Conclusions	44
	Recommendations	44
	Reference	45

List of Tables and Fig

Table No.	Title	Page No.
Table (1-1)	Chemical Composition of Pomegranate	6
Table (1-2)	Phytochemicals of Pomegranate	7
Table (1-3)	General composition of the pomegranate molasses	10
Table (1-4)	Physical and chemical properties of pomegranate molasses.	10
Table (2-1)	Instruments used in this study and their manufacturers.	25
Table (2-2)	Chemicals, used in this study and their suppliers	26
Table (3-1)	Cytogenetic effect of pomegranate molasses (PM) extract and benzoic acid (E210) on mitotic index in human blood lymphocyte culture (in vitro).	33
Table (3-2)	Effect of pomegranate molasses (PM) extract and benzoic acid (E210) on chromosomal aberrations in lymphocytes culture.	34
Table (3-3)	Effect of pomegranate molasses (PM) extract and benzoic acid (E210) on Micronucleus (MN) in lymphocytes culture.	35
Table (3-4)	Effect of pomegranate molasses (PM) extract and benzoic acid (E210) on replication index (RI) in lymphocytes culture	36
Table (3-5)	Interaction between pomegranate molasses (PM) extract and (E-210) on mitotic index in human blood lymphocyte culture (in vitro)	39
Table (3-6)	The effect of interaction between pomegranate molasses extract and benzoic acid (E210) on chromosomal aberrations (CA) in human blood lymphocyte culture (in vitro)	40
Table (3-7)	The effect of pomegranate molasses extract on benzoic acid (E210) - induced micronuclei formation in human blood lymphocyte culture (in vitro)	41
Table (3-8)	The effect of pomegranate molasses extract ((PM)) on replication index (RI) in human lymphocytes induced by benzoic acid (E210)	41
	List of Fig	
Fig(1-1)	*Punica granatum* L(Puniacaceae	4
Fig(2-1)	pomegranate molasses	27
Fig(2-2)	Benzoic acid crystals	27

Abbreviation

RPMI-1640	Rosswell Park Memorial Institute 1640
FDA	Food Drug and Administration
CCl4	Carbon tetrachloride
WHO	World Health Organization
DZA	Diazinon
CHD	coronary heart disease
LSD	least significant difference
b.w	body weight
E Number	European Union
CAS	Chemical Abstracts Service
PBS	Phosphate Buffer Slaine
PHA	Phytoheamatoagglutinin
ROS	Reactive Oxygen Species
MN	Micronucles
MI	Mitotic index
SCE	Sister Chromatid Exchange
CA	Chromosomal Aberration
HMF	hydroxymethylfurfural
DMSO	Dimethyle Sulphoxide
EINECS Number	European Inventory of Existing Commercial chemical Substances
ADI	Acceptable Daily Intake
PJE	Pomegranate juice extracts
RI	Replication Index
PM	pomegranate molasses
EGCG	epigallocatechin gallate

RDA	Recommended daily Allowance
GAE	gallic acid
OP	Organo phosphorus
CHD	coronary heart disease
PCP	Pentachlorophenol
BHA	butylated hydroxyanisole
BHT	butylated hydroxytoluene
ADHD	attention deficit hyperactivity disorder
CHD	coronary heart disease
EFSA	European Food Safety Authority
SMART	somatic mutation recombination test
PHA	Phytohaemagglutinin

Abstract

Due to the lack of other data in relation to the genotoxic effect of pomegranate molasses Therefore, This study came to investigate the role of pomegranate molasses in the inhibition of cytogenetic effect of benzoic acid (E210) on human blood lymphocyte using mitotic index (MI), chromosomal aberration assay (CAs), replication index (RI) and micronucleus test (MN), as parameters, also we evaluated the cytogenetic effect of different concentrations (5,10, and 15 µg / ml) of the molasses, with comparison of the effect of benzoic acid on human blood lymphocyte. Further, we studied the protective effect of pomegranate molasses, included three types of treatment (pre , post- and with benzoic acid (E210)) in order to determine the mechanisms of this product in preventing or reducing the cytogenetic effect of benzoic acid (E210). The results suggest that the pomegranate molasses in concentrations used, does not have genotoxic potential, while the benzoic acid (E-210) has a toxic effect and it is substantiated by an increase in chromosomal aberrations (CA), micronucleus frequency, decreased mitotic and replication index activity in peripheral blood lymphocytes, compared with negative control. Therefore pomegranate molasses is considered as desmutagen in the first order and bio-antimutagen in the second order.

Keywords: cytogenetic, pomegranate molasses, benzoic acid (E-210), peripheral human blood lymphocytes.

Introduction

The continuous increase of the world's population requires the identification of new food resources, as well as the finding of effective methods to ensure their storage and the long-term maintaining and optimization of their nutritional quality, the flavour and the appearance. More than 3,000 food additives play a vital role in today's food supply. A food additive is any substance or mixture of substances other than basic food components, added to food in a scientifically controlled amount (Mpountoukas *et al.* 2008). These food additives (preservatives, antioxidants, emulsifiers, stabilizers, coloring agents, sweeteners, taste and smell improvers, thickeners, gelling agents, anti-caking agents etc.) are used in modern food processing and technology (Sasaki *et al.* 2002).. About 75% of the western diet made up of processed foods (Zengin *et al.*, 2011). So, humans are unavoidably exposed to these complex mixtures in their diet. Some food additives have been prohibited from use due to their toxicity. many studies revealed the potential genotoxic and mutagenic effects of the additives (Gultekin *et al.*, 2013;Martyn *et al.*, 2013 ;Yılmaz et al. 2008, 2009; Mamur *et al.* 2010; Zengin *et al.* 2011).

Potential health risks of food additive are often associated with preservatives which added to food to stop or greatly slow down its spoilage that may leads to loss of quality, edibility or nutritive value that caused by yeasts, molds and bacteria, but also prevent the formation of toxins, especially those produced by bacteria and molds (Altug˘, 2003). The most widely used preservatives is benzoate (benzoic acid, sodium and potassium benzoate) .

Benzoic acid (E210) is commonly used as an antimicrobial substance in many food products, (Sarıkaya and Solak 2003). However, there are some studies that showed the genotoxicity of Benzoic acid in different tests (Yılmaz *et a*l. 2008; Mpountoukas *et al.* 2008). such as Ames tests (Ishidate *et al.*1984; Zeiger *et al.* 1988). Also , it increased the somatic mutations in Drosophila Smart test (Sarıkaya and Solak 2003). Yılmaz *et al.* (2009) reported that benzoic acid significantly increase the chromosomal aberrations and decrease the mitotic index in human lymphocytes.

There has been an enormous interest worldwide in nutraceuticals, which are known to play a pivotal role in health management. Many different studies have shown the beneficial effects of a range of different fruits, vegetables, and spices. Dietary fibers are of increase importance in chronic diseases, and reinforcing their place in the diet, to face these true 'epidemics' (Asif,2011). Therefore , it is prefered the use of industrial products are based on natural sources,like vegetables and fruit (which are rich in phytonutrients and antioxidants), which they are an important factors influencing human health. Epidemiological evidences suggested that regular consumption of fruits and vegetables may reduce the risk of some diseases, including cancer (Liu.,2004).

Pomegranate is commonly eaten around the world and has been used in folk medicine for a wide variety of therapeutic purposes (Shabtay et al.,2008; Ross., 2009). Pomegranate and its products has been reported to be a rich source of bioactive polyphenols, compounds with antioxidant properties such as anthocyanins, hydrolysable and condensed tannins (Sudheesh and Vijayalakshmi, 2005; Lansky and Newman, 2007; Wang et al., 2010). Pomegranate juice extracts (PJE) have been shown to inhibit cellular proliferation and tumor growth and induce cell death via apoptosis in a number of cancer cell lines (Dahlawi et al.,2012).

Natheer et al. (2014) reported that pomegranate granatum seed extract, especially at the third dose (50 mg/kg b.w.) exhibited well protective and anticlastogenic effect against the genotoxic actions of the ifosfamide (is a highly effective chemotherapeutic agent for treatment of a variety of pediatric and adult solid tumors) in mice, using chromosomal aberration test in bone marrow cell. Moreover, Valadares et al. (2010) demonstrated that pomegranate alcoholic extract exerts antimutagenic activity against cyclophosphamide-induced DNA damage and similar findings were shown in the hepatic (Pirinccioglu et al., 2012) and renal (Moneim, El-Khadragy, 2012) tissue of rats pretreated with pomegranate juice and exposed to Carbon tetrachloride (CCl4).

Ávila *et al.* (2013) also showed that the chemoprotective effects of *Punica granatum* L. (Punicaceae) fruits alcoholic extract a protective effect against Cr(VI)-induced genotoxicity on mice.

Pomegranate molasses (PM) is a concentrated product produced simply by boiling, without addition of sugar or other additives (Poyrazoglu *et al.*, 2002; Incedayi *et al.*, 2010), and it is widely consumed in the Middle East; it is rich in more efficient antioxidants than that found in the pomegranate juice (Chalfoun-Mounayar *et al.*, 2012).

Abd Elmonem (2014) had proved the protective effects of pomegranate molasses on the damage induced by Diazinon (DZA) it is one of the organophosphorus (OP) insecticides. The data showed that a high dose of DZN can cause hematological and biochemical changes and the antioxidants PM ameliorate most changes.

Although some studies have addressed the antioxidant effects of the product pomegranate molasses (PM) however, there are no reports in the literatures to date regard the cytogenetic effects and of this product, therefore, our study was aimed to the:

1. Investigate of cytogenetic effects of pomegranate molasses (PM) on human peripheral blood lymphocytes.
2. Investigate the protective effects of pomegranate molasses (PM) extract on modulating the genotoxic effects induced by food additive benzoic acid (E210) on human peripheral blood lymphocytes, by using different parameter :
- Chromosomal Aberrations (CA)
- Micronucleus (MN)
- Mitotic Index (MI)
- Replication Index (RI)

1- Literature Review

1.1 Pomegranate

- **Botanical Name**....... *Punica granatum* L
- **Family Name** .... Puniacaceae
- **Common Name**........ Pomegranate, Anar
- **Part Used** .... Seeds, flowers

Fig(1- 1) *Punica granatum* L(Puniacaceae)

1.1.1 Description

Pomegranate (*Punica granatum* L.) is considered as one of the oldest known edible fruit that is mentioned in the Quran, the Bible, the Torah, and the Babylonian Talmud as 'Food of Gods' that is symbolic of plentiness, fertility and prosperity (Aviram *et al.*, 2000; Seeram *et al.*, 2006) .The pomegranate tree typically grows to 12-16 feet, has many spiny branches, and can be extremely long lived, as evidenced by trees at Versailles, France, known to be over 200 years old. The leaves are glossy and lance shaped, and the bark of the tree turns gray as the tree ages. The flowers are large, red, white, or variegated and have a tubular calyx that eventually becomes the fruit [Figure 1-1]. The ripe pomegranate fruit can be up to five inches wide with a deep red, leathery skin, is grenade-shaped, and crowned by the pointed calyx. The fruit contains many seeds (arils) separated by white, membranous pericarp, and each is surrounded by small amounts of tart, red juice (Jurenka,2008).

1.1.2 Habitat

Native: Afghanistan, Iran, Libia ,Tunisia.
Exotic: Egypt, Greece, India, Indonesia, Italy, Morocco, Russia , Saudi Arabia, Spain, Turkey, United States of America (Langley,2000).

1.1.3 Traditional use

Pomegranate has been used extensively in the folk medicine of many cultures as a "healing food" as a source of traditional remedies for thousands of years (Ross,2009 ; Shabtay *et al* .,2008). Studies had shown that pomegranate has many potential effects including: bactericidal, antifungal, antiviral, immune modulation, vermifuge, refrigerant, astringent, styptic, laxative, diuretic and anthelmintic effect. Moreover, it serves to decrease symptoms effects of cardiovascular diseases, diabetes, diarrhea, dysentery, asthma, bronchitis, cough, bleeding disorders, fever, inflammation, acquired immune deficiency syndrome, dyspepsia, ulcers, mouth lesions, skin lesions, malaria, prostate cancer, atherosclerosis, hypertension, hyperlipidemia, male infertility, alzheimer, obesity and infant brain ischemia (Lansky & Newman ., 2007, Reddy *et al.*, 2007).

1.1.4 Chemical composition of pomegranates

The pomegranate tree can be divided into several anatomical compartments: seed, juice, peel, leaf, flower and root bark, each of which is widely used in therapeutic and food formulas, some studies reported that even the bark, roots, and leaves of these trees have medicinal benefit. Considerable progress has been made in the last decade in establishing the pharmacological mechanisms of pomegranate and its individual constituents. Pomegranate juice contains anthocyanins, glucose, ascorbic acid, ellagic acid, gallic acid, caffeic acid, catechin, epigallocatechin gallate (EGCG), querticin, rutin, iron, and amino acids. Pomegranate seed oil is composed primarily of punicic acid and sterols. The pericarp (peel, rind) contains punicalgins, flavones, flavonones, and other flavanols. Tannins, including punicalin and punicafolin, and flavone glycosides like luteolin and apigenin, form important constituents of pomegranate leaves. The flowers of pomegranate are composed of ursolic acid, triterpinoids like maslinic acid, and asiatic acid. Ellagitannins and piperidine alkaloids are present in pomegranate roots and bark which are good source of tannins, dyes(Khan, 2009; Viuda-Martos *et al.*, 2010; Wang *et al.*, 2010). However, current research seems indicate that the most therapeutically beneficial pomegranate constituents is due to ellagic acid ellagitannins (including punicalagins), punicic acid, flavonoids anthocyanidins, anthocyanins, and estrogenic flavones.

Pomegranate aril (an extra seed-covering), juice provides about 16% of an adult's daily vitamin C requirement per 100 ml serving, and is a good source of vitamin B5 (pantothenic acid), potassium, and natural phenols, such as ellagitannins and flavonoids (Tiwari ,2012).

The chemical composition of the pomegranate and its products depends on the cultivar, growing region, and climate, the fruits stage of maturity, cultural practices and manufacturing systems (Dumas *et al.*, 2003; Raffo *et al.*, 2006, Borochov-Neori *et al.*, 2009; Zarei *et al.*, 2011). The chemical composition of pomegranate fruit and phytochemicals in pomegranate tree parts are show in tables (1-1) and (1-2).

Table 1.1 Chemical Composition of Pomegranate* (Yilmaz, 2007).

Constituent	Percentage %
Moisture	72.6-86.4%
Protein	0.05-1.6%
Fat	0.01-0.9%
Mineral elements	0.36-0.73%
Fibre	3.4-5.0%
Carbohydrates	15.4-19.6%
Calcium	3.0-12.0 mg
Phosphorus	8.0-37.0 mg
Iron	0.3-1.2 mg
Sodium	3.0 mg
Magnesium	9.0 mg
Ascorbic acid (Vitamin C)	4.0-14.0 mg
Thiamine	0.01 mg
Riboflavine (Vitamin B2)	0.012-0.03 mg
Niacine	0.18-0.3 mg

*Values per 100 g of edible portions

Table1.2 Phytochemicals of Pomegranate (Jurenka, 2008)

Plant Component	Constituents
Pomegranate juice	Anthocyanins; glucose; ascorbic acid; phenolics such as ellagic acid, gallic acid, caffeic acid, catechin, epigallocatechin gallate (EGCG), quercetin, rutin; mineral elements; amino acids
Pomegranate seed oil	Punicic acid; ellagic acid; fatty acids; sterols
Pomegranate pericarp (peel, rind)	Phenolic compounds like punicalagins, gallic acid, catechin, EGCG, quercetin, rutin, anthocyanidins, other flavonoids
Pomegranate leaves	Ellagitannins (punicalin and punicafolin); flavonols such as luteolin and apgenin
Pomegranate flower	Gallic acid, triterpenoids such as ursolic, maslinic and asiatic acid
Pomegranate roots and bark	Ellagitannins; piperidine alkaloids

1.2 Pomegranate fruit derived products

Pomegranate can be consumed as fresh, fruit juice, fermented fruit juice, dried aril, frozen aril, minimally-processed aril, canned aril, jam, jelly, wine, vinegar, paste, fruit leather and in flavoring products.

Pomegranate arils can either be consumed fresh or procesed (dried, frozen, canned and minimally-processed) (Al-Maiman and Ahmad, 2002). The dehydrated arils are acidic (7.8–15.4%), help in improving taste and digestion, and are widely used as acidulent in culinary preparations. The dried anardana contains acid (5.8-15.4%), total sugars (9.3-17.5%) and crude fiber as compared to fresh fruit. To obtain frozen arils the arils are put into polyethylene bags either with syrup of 15° Brix or coated with solid sugar and frozen in a chest freezer. For canned arils, used generally as an appartiser, the arils were put into metal tins with syrup of 15° Brix and sterilised for 10 minutes. In the production of minimally processed pomegranate aril pomegranates are chilled to 0°C, selected, washed and dried with a current of air at room temperature. They are conditioned in polyethylene bags that were heat-sealed and conserved in a chamber at 0°C for 10-15 days. These arils are used as a garnish for desserts and salad (Al-Maiman & Ahmad, 2002).

Pomegranate juice can be extracted by using a spiral-type screw press without crushing the seeds. The juice is clarified by heating in a flash pasteurizer at 79-82°C they cooled, settled for 24 hours and filtered. The clear juice can be preserved by heat treatment or by using chemicals. The use of sulphur dioxide is banned for pomegranate due to loss of colour by bleaching action of SO2. Pomegranate juice represents one of the foods recently promoted for its health benefits since a glass of pomegranate juice contains about 40% of the Recommended daily Allowance (RDA) of Vitamin C (Singh & Singh, 2004).

Pomegrenate syrup of 60° Brix with an added acidity of 1.5% as citric acid has a bright purplish-red colour and a delightful taste and flavour. It was preserved by pasteurization. Preparation of jelly on a small-scale from sweet-sour pomegranates is described by Adsule *et al.* (1992) and Singh & Singh (2004). When making the jellies, approximately 50% of the total anthocyanins present in the juice of are lost. During storage at 5°C, certain colour differences were observed, which indicates that the pH was not the only parameter responsible for this characteristic. For preparation of wine, the whole furits are pressed without crushing or juice may be extracted from pomegranate grains, which gives a yield of 76 to 85% (Adsule & Patil, 1995). Sugar is added to the juice to obtain 22-23° Brix. The juice is fermented as in the same manner of red grape wine.

The wine is flash pasteurized at 60°C and bottled hot (Singh & Singh, 2004). Pomegranate seed is a residue obtained from pomegranate juice production, ranging between 40 and 100 g/kg of fruit weight (Fadavi *et al.*, 2006; Lansky & Newman, 2007).

The seeds are rich source of lipids, and the fatty acid component of pomegranate seed oil comprises over 95% of the oil, of which 99% is tri acylglycerols. Minor components of the oil include vitamin E, sterols, steroids, and a key component of mammalian myelin sheaths, cerebroside (Tsuyuki *et al.*, 1981). There are different market products of Pomegranate in Turkish:

'Anardana' which constitute a traditional product it is lies in the drying seeds along with pulp (arils) (Pruthi & Saxena, 1984).

'solid pekmez',which has a pasty form that is easily spread on a slice of bread. The 'pomegranate leather (pestil)' is another Turkish pomegranate derived product that can be stored for a long time without deterioration. Pomegranates are washed, granulated, crushed, pressed and filtered to separate the seeds and skin. Pekmez earth is added to neutralize and clarify the fresh pomegranate juice. Clarified juice is filtered and is mixed with the wheat starch. Nuts such as walnut or hazelnut can be added in small pieces if desired. The juice and starch mixture is concentrated upto 40° Brix by boiling and continuous stirring. The puree is spread on cloths of 0.5–2.00 mm thickness and sun-dried until a mild, tasty, light and chewable leathery product is obtained. The dried pestil is folded, cut and stored in dry conditions (Maskan *et al.*, 2002). In recent years, pomegranate has become a widely known product. The production growth, consumption and trading volume have been increased because of the significant developments in growing techniques, nutritional technologies, storage and transportation means (Vardin and Abbasoğlu, 2004). The most important of these products is pomegranate molasses

1.3 Pomegranate molasses

There are different market brands of pomegranate molasses purchased from the local markets of Bursa, Turkey it is called (sour pomegranate pekmez, nar eksisi, pomegranate sauce)', a traditional seasoning commonly used in salads and many dishes to improve the taste and aroma of food. It is a concentrated product produced simply by boiling, without the addition of further sugar or other additives (Poyrazoglu *et al.*, 2002; Incedayi *et al.*, 2010). Pomegranate molasses is a highly nutritive product since it is more concentrate and have a high mineral content. Traditional methods are still being used to produce pomegranate molasses, of which requires cleaning, crushing, extraction, filtration, and evaporation (up to 35-65° Brix) in an open vessel or under vacuum. Clarification is not recommended in pomegranate molasses since customers prefer bitterness and sourness that comes from phenolic substances and acidity (Vardin & Abbasoglu, 2004; Kaya & Sozer, 2005).

General composition physical and chemical properties of the pomegranate molasses are illustrated in Tables 1-3 and 1- 4 respectively

Table 1. 3. General composition of the pomegranate molasses (Anonymous,2001)

Components for each (100) ml	Amount
Water Soluble Dry Matter, minimum	68.0%
Titration acidity (as citric acid), minimum	7.5%
pH	3.0%
*HMF, mg/100g, maximum	50%
Saccharose	Not allowed

*HMF: hydroxymethylfurfural

Table 1.4. Physical and chemical properties of pomegranate molasses.(Fadavi *et al.*2005; Yilmaz *et al.*2007).

Property for each (100) ml	Value
Moisture (%)	24.43±2.12
Brix	73.90±2.30
Protein (%)	0.23±0.06
Ash (%)	Trace
calcium	280.30±15.10
phosphorus	15.57±1.97
potassium	20.12±0.20
iron	15.81±1.96
zinc	7.14±1.70
magnesium	28.00±0.59
Copper	0.61±0.01
Total phenolic content (mg GAE*/g)	52.56±20.68
Color values (Hunter Lab)L	1.88±0.52

*GAE : gallic acid
*Values per 100 g of edible portions

1.3.1 Antioxidant activity

Over the past few years, consumer demand-based research on functional foods gave a basis for traditional using of pomegranate, which lead to an increase in number of scientific papers concerning pomegranate and its products with health-improving effects (Turk *et al.*, 2008; Dai *et al.*, 2010; Jadeja *et al.*, 2010; Park *et al.*, 2010). The reports have focused on in vitro, ex vivo, and in vivo antioxidant actions of pomegranate and its products, and reactive oxygen species (ROS) elimination and inhibit ROS generation. Oxidative stress, refers to a cell state characterized by excessive production of ROS, and it has been given growing attention, as the generation of ROS, causing improved oxidative stress. That induce DNA damage and trigger redox-dependent transcription factors which lead to cancer, aging inflammatory, cardiovascular and neurodegenerative diseases (Kryston et al., 2011; Martin et al., 2011).

The effect of pomegranate cultivars on antioxidant activity was target study by some authors (Pande & Akoh, 2009; Sadeghi et al. 2009). All authors reported considerable variation in some of the chemical composition profile (lipids, phenols, organic acids, vitamins, sugars) and antioxidant properties of pomegranate different samples, independent on the antioxidant method performed. The antioxidant activity of pomegranate and its products was almost determined via in vitro trials and several methods could be used for its determination, however, pomegranate showed an antioxidant activity (independent on the antioxidant test) with significant linear correlation with phenolic content (Elfalleh et al., 2009).

Seeram et al. (2005) stated that the antioxidant level in pomegranate juice was higher than found in other fruit juices, such as blueberry, cranberry, and orange. Guo et al. (2008) and Gil et al. (2000) demonstrated that pomegranate juice and seed extracts have 2-3 times the in vitro antioxidant capacity of either red wine or green tea. Several works had demonstrated that peel and seeds have antioxidant activity (Sumathy et al.2013).

Chalfoun-Mounayar et al (2012) studied the antioxidant activities of pomegranate molasses and pomegranate juice, they found that pomegranate molasses possesses more efficient antioxidants than pomegranate juice, as well, Abd- Elmonem (2014) proved the protective effects of pomegranate molasses (PM) on the hematological and biochemical damages induced by high dose of Diazinon (DZA), it is one of the organo phosphorus (OP) insecticides , in the same way, West et al (2007) showed that polyphenols present in this plant protect neonatal mouse brain against hypoxic-ischemic injury.

1.3.2 The health benefits of pomegranate derived products.

It is clear that bioactive compounds present in daily diet, mainly in fruits and vegetables, have prevention potential in cancer through inhibiting carcinogenesis via cell-defensive and cell-death mechanism regulation. These chemopreventive effects may be attributed to a complex effect of various phenolic substances of antioxidant capacity (Khan et al., 2008). Pomegranate is rich in anthocyanins, 3-glucosides, 3,5-diglucosides of delphinidin, cyanidin and pelargonidin, ellagitannins and other phenolic compounds, which are known bioactive compounds with antioxidant and antitumor activity (Chaturvedula et al., 2011; Zhang et al., 2011). Major hydrolysable tannins in pomegranates are gallo tannins, ellagic acid tannins and gallagyl tannins, generally termed as punicalagins, and they have been shown to inhibit the proliferation of human cancer cells and modulate inflammatory subcellular signaling pathways due to a high antioxidant activity (Seeram et al., 2005). There are several studies conducted to evaluate the efficacy of pomegranate and its products as an anti-proliferative, anti-invasive, and pro-apoptotic agent in various cancer cell lines such as skin, prostate, breast, colon, and blood cancer (Dahlawi et al.,2012).

Adams et al. (2006) revealed that pomegranate juice suppresses cancer activity through the combined antioxidant and antiinflammatory effects by modulating inflammatory cell signaling in colon cancer cells. Paller et al. (2012) suggested that pomegranate juice may have cancer chemopreventive effect as well as cancer-chemotherapeutic effects against prostate cancer in humans. Pomegranate fruit extract possesses aremarkable antitumor effects on mouse skin.

Researchers found that daily consumption of pomegranate juice may stress induced myocardial ischemia sump toms in patients who have coronary heart disease (CHD) and preventes hardening of the arteries by reducing blood vessel damage, but also reversed the progression of CHD (Sumner et al., 2005). Hartman et al. (2006) reported that pomegranate juice had beneficial effect in animal model of Alzheimer's disease since polyphenols are responsible for neural protection. Natheer et al . (2014) reported that the P. granatum seed extract, especially at the third dose (50 mg/kg b.w.) exhibited well protective and anti-clastogenetic effect against the genotoxic actions of the ifosfamide (is a nitrogen mustard alkylating agent used in the treatment of cancer) in mice, using chromosomal aberrations in bone marrow cell test. Valadares et al (2010) demonstrated that pomegranate alcoholic extract exerts antimutagenic activity against cyclophosphamide-induced DNA damage, similar findings were shown in the hepatic (Pirinccioglu et al., 2012) and renal tissue of rats pretreated with pomegranate juice and exposed to Carbon tetrachloride (CCl4) (Moneim &El-Khadragy, 2012). Ávila *et al.* (2013) also showed that the chemoprotective effects of Punica granatum alcoholic fruits exerts a protective effect against Cr(VI)-induced genotoxicity on mice ,and other study have shown the ability of pomegranate peels extract to scavenge various reactive oxygen species and inhibit lipid per oxidation, and the peels extract protect against Pentachlorophenol (PCP) induced oxidative stress, cytogenetic toxicity and hepatic injury (Fatma *et al* ,2013)

1.4 Food additives

With the increase in the production of processed and convenience foods, food additives have become an increasingly important practice in modern food technology (Saad *et al.*, 2005). Additives are widely used for various purposes, including preservation, coloring, and sweetening (Cheng *et al.*, 2010; Lerner and Lerner 2011; Becerril *et al.*, 2013). A food additive is any substance added to food. Legally, the term refers to any substance the intended use which results directly or indirectly-in becoming a component or affecting the characteristics of any food. This definition includes any substance used in the production, processing, treatment, packaging, transportation or storage of food (Kunkel and Barbara., 2004). Food additives are also substances added to food to preserve flavor or enhance its taste and appearance. Some additives have been used for centuries; for example, preserving food by pickling (with vinegar), salting, or using sulfur dioxide in some wines as sweets. With the advent of processed foods in the second half of the 20th century, many more additives have been introduced, of both natural and artificial origin (Food and Drug Administration (FDA). 2013). Direct additives are those that are intentionally added to foods for a specific purpose while indirect additives are those to which the food is exposed during processing, packaging, or storing. Additives also improve the nutritional value of certain foods and can make them more appealing by improving their taste, texture, consistency or colour (Houghton.,2002). Food additives are used to increase the nutritional value or to protect decaying nutrients during the preparation of factory-made foods (Greenwald *et al,* 2001; Bezar *et al*, 2002).

1.4.1 Classification of food additives

Food additives can be divided into several groups, although there is some overlap between them (Watson *et al.*,2002; Czarra.,2009 ; Sarikaya *et al.*, 2012).

- Acids: Food acids are added to make flavors "sharper", and also act as preservatives and antioxidants. Common food acids are vinegar, citric acid, tartaric acid, malic acid ,fumaric acid, and lactic acid
- Acidity regulators: they are used to change or otherwise control the acidity and alkalinity of foods.
- Anticaking agents: they keep powders such as milk powder from caking or sticking.
- Anti foaming agents: reduce or prevent foaming in foods.
- Antioxidants: prevent rancidity in fatty foods and damage to foods caused by oxygen. Examples of antioxidants vitamin C, vitamin E, BHA (butylated hydroxyanisole), BHT (butylated hydroxytoluene), and propylgallate.
- Bulking agents: such as starch are additives that increase the bulk of a food without affecting its nutritional value.
- Food coloring: added to food to replace colors lost during preparation, or to make food look more attractive and appealing.
- Retention agents: are used to preserve a food's existing color.
- Flavors: are additives that give food a particular taste or smell, and may be derived from natural ingredients or created artificially.
- Flavor enhancers: they enhance a food's existing flavors. They may be extracted from natural sources (through distillation, solvent extraction, maceration, etc) or created artificially.
- Flour treatment agents: are added to flour to improve its color or its use in baking.
- Glazing agents: provide a shiny appearance or protective coating to foods.
- Humectants: prevent foods from drying out.
- Tracer gas: allows for package integrity testing preventing foods from being exposed to atmosphere, thus guaranteeing shelf life.
- Stabilizers: Stabilizers, thickeners and gelling agents, like agar or pectin (used in jam for example) give foods a firmer texture, while they are not a true emulsifiers, they help to stabilize emulsions .

- Sweeteners: added to foods for flavoring. Sweeteners other than sugar are added to keep the food energy (calories) low, or because they have beneficial effects for diabetes mellitus tooth decay and diarrhea.
- Thickeners: are substances which when added to the mixture increase its viscosity without substantially modifying its other properties.

1.4.1.1 Preservatives

This group is consists of antimicrobials, antioxidants, and anti browning agents. The **E** numbers (codes for substances that can be used as food additives for use within the European, They are commonly found on food labels throughout the European, Union Safety assessment and approval are the responsibility of the European Food Safety Authority) of the preservatives range from E200 -E399.

The preservatives are added to stop or delay nutritional losses due to microbiological, enzymatic or chemical changes of foods and to prolong the shelf life and quality of foods it slow down its spoilage polluted which is caused by microorganisms present in the package or the hands that may deal with before packaging . Although preservatives are essential to maintain food safety, but some modern synthetic preservatives have become controversial because they have been shown to cause respiratory problems, different allergies, attention deficit hyperactivity disorder (ADHD) in some people who are sensitive to specific chemicals and other health problem (Louis and Botulism., 1991).

1.4.1.2 Antimicrobials

Antimicrobials are added to food for 2 purposes:-
(1) to control natural spoilage of food (food control) and/or
(2) to avoid / control contamination by microorganisms, including pathogenic ones (of food safety concern) (Tajkarimi *et al.*, 2010).

The main antimicrobials used in food with quantum satis status are acetic acid (E260), potassium acetate (E261), calcium acetate (E263), lactic acid (E270), carbon dioxide (E209), and malic acid (E296). The antimicrobial additives with restricted uses are benzoic acid and benzoates (E210-E219; ADI 5 mg/kg bw), sorbic acid, and sorbates

(E200-E209; ADI 25 mg/kg bw), propionic acid and propionates (E280-E289; quantum satis), nitrites (potassium nitrite E249; ADI 0.07 mg/kg bw, sodium nitrite E250; ADI 0.1 mg/kg bw). Nitrates (sodium nitrate E251 and potassium nitrate E252; both with ADI 3.7mg/kg bw), and parabens (E214-E219; ADI 10 mg/kg bw).

Most studies suggest the negative effect of antimicrobials. For example sulfites are a group of molecules used in foods as antimicrobials and antibrowning agents(the most common are sulfur dioxide, sodium and potassium bisulfite, and, sodium and potassium bisulfate) . Their antimicrobial effect is carried out by the uptake of SH groups from sulfites into the microorganism's cell where they react with proteins, DNA, enzymes, that the antioxidant effect occurs by inhibiting both maillard reactions and the enzyme polyphenol oxidase. Sulfites can act freely or combined with organic acids, being used in wine making and in many other foodstuff that is prone to microbiological decay. The negative effects of sulfites are related with the destruction of vitamin B1 (thiamine) and to cause skin and respiratory sensitivities, such as dermatitis, urticaria, angioedema, abdominal pain, diarrhea, bronchoconstriction, and fatal anaphylaxis (Rencuzogullari *et al*., 2001; Valley *et al*., 2009; Garcıa-Gavın *et al* ., 2012). These symptoms can become more prevalent due to the large quantity of foodstuffs treated with sulfites, like canned goods, seafood, and dried fruits. In the EU and United States, all products that contain sulfites must show this information on the label.

Many surveys have been carried out in various countries. Belgium and New Zealand reporting that it is danger to individuals who consume high amount of sulfited wine (Cressey and Jones 2009; Vandevijvere *et al.,* 2010). In Turkey, some foodstuffs intended for export displayed quantities of sulfites above the legal limit, while Korea reported that the consumption of sulfites for individuals aged between 30 and 64 is above that of the other age groups (Suh *et al.,* 2007; Ulca *et al* ., 2011). More research should be carried out to determine the real effect of sulfites to human health, especially when used as a food additive.

Propanoic or propanoic acid (E280) is a naturally occurring carboxylic acid that is used in food, especially bakery products, to avoid mold and other fungal contamination. There are not many studies regarding the toxicity of propionic acid or its salts, (sodium propionate, E281), calcium propionate (E282), and potassium propionate (E283), although it has been considered to suppress, in a dose-dependent manner, T-helper –type1 immune response in human peripheral blood mononuclear cells in vitro. Sodium propionate has been stated as inducing abnormalities on the root tips of onion (Maier *et al* ., 2010), while calcium propionate has been related to irritability, restlessness, inattention, and sleep disturbance in some children (Dengate and Ruben.,2002;Turkoglu.,2008).

Nitrates (E240-E259) and nitrites (E249, E250) in which sodium nitrate (E251), potassium nitrate (E252), potassium nitrite, and sodium nitrite are the most important compounds, they are used in the meat industry, mainly for curing. While nitrate was widely used in the past, nowadays it is restricted to specific slow meat curing. On the other hand, nitrites are used for various applications in several types of meat, mainly for color formation, flavor enhancement, and antimicrobial activity. Nowadays, nitrites are considered the only food additive that can inhibit the development of the botulinum toxin, thus justifying their use in a benefit/risk scale in food industry. The EFSA allows its use at the minimum possible dosage. Apart from being used as food preservatives, nitrites are also present in considerable quantities in non-treated vegetables and fruits. These compounds are also known to take part in the formation of nitrosamines (carcinogenic molecules resulting from the reaction of nitrites with secondary amines) posing a threat to consumers (Sebranek and Bacus2007; Sindelar and Milkowski, 2012). Nitrites are considered by some authors as being carcinogenic, while others refute this possibility and consider plant nitrites important for some physiological roles, such as supporting cardiovascular health and gastrointestinal immune function (Hord *et al* .,2009). Although evidence supports both theories, it is widely accepted that the excess intake of nitrite is dangerous and has deleterious effects on human health, (Cammack *et al.,* 1999). In order to counter these adverse effects, much research is being carried out to find alternatives to nitrites (Chan *et al.,*2011; Hord.,2011).

Paraben is a generic name for a group of food additives, which are alkyl esters of p-hydroxybenzoic acid. These compounds are widely used in food as antimicrobials, especially due to their absence of odor or/and taste (Tavares et al., 2009). Its effectiveness increases as a function of the alkyl group length. The most used compounds of this group are methyl paraben (E218; ADI 10 mg/kg bw), ethyl paraben (E214; ADI 10 mg/kg bw), and propyl paraben (E216; ADI 10 mg/kg bw). In the past, parabens were not considered mutagenic, but were known to cause chromosomal aberrations and contact allergy (Darbre et al., 2004; Tavares et al., 2009).

1.4.2 Effect of additives in food

There are several researches concerning food additives (Patel et al., 2010; Maqsood et al., 2013), their impact on health and behavior (Gultekin et al., 2013;Martyn et al., 2013), as well as the methods to detect them (Frazier et al., 2000; Saad et al., 2005; Wang et al., 2006; Garcıa-Jimenez and Capitan-Vallvey,2007; Yoshioka and Ichihashi 2008; Cantarelli et al., 2009; Merusi et al., 2010; Yoshikawa et al., 2011; Ohtsuki et al., 2012; Pundir and Rawal., 2013) . Food additives were considered desirable due to the function they carried out, but could, in some cases, have negative effects on health (Bronzwaer 2008). Food Safety Authority does not allow the use of additives that are associated with increased risk of cancer at the allowed doses that they are used in foods (Greenwald et al, 2001; Bezar et al, 2002), therefore some food additives have been prohibited from use due to their toxicity. For example, AF-2 (2-[2-furyl]-3-[5-nitro-2-furyl] acrylamide) which is formed during food heating , is proved to induce DNA damage and mutation in bacteria insects and mammalian cells in vivo and in vitro (Alzahrani et al .,2011).

1.5 Benzoic acid

Benzoic Acid

	Benzoic acid 99%
Molecular weight:	122.12
Molecular Formula:	$C_7H_6O_2$
Purity:	99%
color:	white
CAS No.	65-85-0
Appearance:	White flakes or granules

1.5.1 Effect of Benzoic acid

Benzoic acid (E210), Benzyl alcohol and its derivatives (benzoic acid and sodium benzoate, among others) belong to the aryl alkyl alcohols, which apart from being food additives are also commonly used in fragrances, cosmetics, shampoos, soaps, and other toiletries as well as in household cleaners and detergents (Scognamiglio *et al.* 2012). Benzoic acid (E210), is commonly used as an antimicrobial substance in many food products (Sarikaya and Solak, 2003), employed against yeast, bacteria, and fungi. It acts through membrane disruption and inhibition of metabolic reactions, stress, and accumulation of toxic anions inside the microbial cell (Bruel and Coote 1999). However,There are some studies that showed the genotoxicity of benzoic acid in different tests (Yılmaz *et al.* 2008; Mpountoukas *et al.* 2008). such as Ames tests (Ishidate *et al.*1984; Zeiger *et al.* 1988). and also , it Caused increased the somatic mutations in somatic mutation recombination test (SMART) (Sarıkaya and Solak 2003). While, Yilmaz *et al.* (2009) reported that benzoic acid significantly increased the chromosomal aberrations and decreased the mitotic index in human lymphocytes.

Benzoic acid may be coupled to calcium, potassium, or sodium for different food. The main applications of sodium benzoate (E211) are soft drinks, fruit juices, sauces, pickles, edible coatings, seafood products, toothpastes, lotions, creams, and some pharmaceutical products (WHO

2000). This antimicrobial compound has been tested in vitro, and was regarded as nontoxic, but some authors found toxicity in the drosphila SMART (somatic mutation recombination test) test, root tips of garlic (*Allium sativum*), as well as a clastogenic, mutagenic, and cytotoxic effect in human peripheral blood lymphocytes (Yilmaz *et al.*, 2008, 2009; Zengin *et al.*, 2011). In murine models, reviewed the risk of exposure of benzyl alcohol, benzoic acid, and sodium benzoate and concluded that, although being safe, in some studies regarding mice, malformations and toxicity effects were detected and dermal complications were known to occur in humans. Furthermore, due to the various applications of these compounds, the risk of inhalation could not be determined, and remained as an important and urgent matter for further studied. In a study involving children (3-years old of 8-9-years old) where exposed to sodium benzoate, a global hyperactivity aggregate was reported when compared to the control group. Despite that sodium benzoate was mixed with food colorants, being difficult to determine which of the compounds was the responsible for the hyperactivity or a synergistic effect where existed, but interesting conclusions where found (McCann *et al*., 2007). The same hyperactivity behavior was reported in a study involving college students who consumed sodium benzoate rich soft drinks, validating the need for further studies regarding this compound (Beezhold *et al*., 2014). In Portugal and Italy, surveys were carried out in order to discover if soft drinks contained benzoic acid or benzene residues. The results pointed out that both benzoic and sorbic acids were present above the allowed values in some products, although not exceeding the acceptable daily intakes. Benzoates are used with acidic foods, sorbates are employed to alkaline food in the case of benzene, it was found in some Italian soft drinks (Ulca *et al.*, 2013). Turkish surveys detected both of these antimicrobials in various foods bay above the maximum limit in some food samples of the Turkish Food Codex (Lino and Pena., 2010; Bonaccorsi *et al.*, 2012; Cakir and Cagri-Mehmetoglu., 2013).

Adams *et al.* (2005) reported that the safety of benzyl derivatives in food was supported by the higher intake of those compounds in traditional foods rather than in the intentionally added flavorings. Animal models and human dermatological assays where done regarding absorption of benzyl alcohol and detect the toxicological and dermatological effect of aryl alkyl alcohols used in fragrances, and determined that with all the data gathered, these compounds do not pose safety concerns in the declared levels of exposure (Belsito *et al.* 2012).

Still, very little research focused on occupational exposure to these compounds, and future studies should take into consideration the combined effects of occupational and non occupational (cosmetics, perfumes, shampoos) quantities absorbed by the skin (Schnuch *et al*., 2011). Although these reports seem to be unsettling, benzoates are necessary and the only way to be removed as an additives is when a non toxic substitute with the same effect is found.

1.6 Cytogenetic analyses

Cytogenetic is a branch of genetics that is concerned with the study of the structure and function of the cell, especially the chromosome. Cytogenetic analyses are essential to the diagnosis and treatment of different forms of cancer by the analysis of blood or bone marrow cells that reveals the organization of chromosomes (Chauhan *et al.*,2007).Chromosomal aberrations, mitotic index and micronucleus analysis of human lymphocytes as well as replicative index are the most useful assays to detect the potential genotoxicity of food additives (Yüzbasıoglu *et al.* 2006; Mpountoukas *et al.* 2008; Yilmaz *et al,.*2009; Mamur *et al.* 2010; Zengin *et al.* 2011). The genome damage of the peripheral blood lymphocytes has been widely used as a biomarker of carcinogenesis from genotoxic environmental factors (Hagmar *et al*.2004).

1.6.1 Mitotic index (MI)

The mitotic index is a reliable predictor of cell proliferation in the tissue. The mitotic index assay is used to characterize the proliferating cells and to identify the compounds that inhibit mitotic progression resulting in decrement in the MI of that population. It is estimated as the ratio between the number of cells in mitosis and the total number of the cells. When colchicines or its derivative (colcemid) are added, this can arrest the cell cycle at metaphase (stage) leaving the chromosomes in their visible form. Colchicines disrupt the microtubules formation which are necessary for the spindle fibers to separate the chromosomes during anaphase (Fahad and Salim, 2009).

Mitotic abnormalities often arise directly from defects of centrosome and /or mitotic spindles, which then induce prolonged mitotic arrest or delayed mitotic occurrence and trigger induction of apoptosis (Mollinedo and Gajab, 2003). Recent reports have demonstrated that entry into mitosis in the presence of damaged DNA leads to inactivation of centrosomes, formation of aberrant spindles and blockage of chromosome segregation, which consequently delays mitosis progression (Hut *et al.,* 2003).

In addition, chemical or pharmacological inhibition of the DNA damage checkpoint at the G2 stage induces premature entry into mitosis and subsequent initiation of apoptosis (Sampath and Plunkett, 2001).The research had indicated that this assay was effected by the mutagenic and carcinogenic materials in which all of these materials can affect MI either in vivo or in vitro (Guleray and Lokman, 2005; Ekanem and Osuji, 2006).

1.6.2 Micronucleus assay (MN)

Micronuclei are cytoplasmic chromatin–containing bodies formed from chromosome fragments or chromosome lag during anaphase and fail to become incorporated into daughter cell nuclei during cell division because of a genetic damage that results in chromosome breaks. Structurally abnormal chromosome or a spindle abnormalities leads to micronucleus formation. The incidence of micronuclei serves as an index of these types of damage (Armstrong and Galloway, 1994; Selvendiran et al., 2005). The key advantage of the MN assay is the relative ease of scoring and the statistical power obtained from scoring larger numbers of cells than are typically used for metaphase analysis, when kinetochore or centromere detection methods are used it is possible to distinguish between MN caused by chromosome breakage and MN caused by chromosomal segregation (Fenech, 2007). This assay performed to detects clastogenicity due to chromosome breakage, due to chromosome lagging resulting from dysfunction of mitotic apparatus (Jorge et al., 2003).Thus, the micronucleus assay has been widely used to measure genotoxicity, both in vivo and in vitro (Fenech et al., 2003; Fenech, 2010)

A positive results in the micronucleus test indicate that the investingated substances are micronuclei inducer, which are the result of chromosomal or mitotic apparatus, in the erythroblasts of the test species. A negative results indicate that, the tested substance does not induce chromosomal or spindle damage leading to the formation of micronuclei in the immature erythrocytes of the test species (Kirsch-Volders et al .,2004).

1.6.3 Chromosome aberrations (CA)

During the course of meiosis, portion of chromosome are often relocated, moving within the chromosome itself or between different chromosome. This process produces changes in the morphology of the chromosome itself, which are referred to as chromosomal aberrations (CAs) (Holland, 2005).The type of chromosomal structural that may depend on the lesions induced in the DNA alterations produced by physical and chemical agents. Chromosomal aberrations mostly induced by substances that directly break the backbone of DNA (e.g. ionizing radiation and radiomimetic chemicals) or significantly distort the DNA helix (e.g. inter chelating agents). Structural aberrations can be classified as either unstable or stable, indicating their ability to persist in dividing cell populations (Carrano, 1986).

According to Meera and Nagarjuna, (2010) all chemicals that produce DNA damage leading to mutation or cancer are described as genotoxic. One of the best ways to minimize the effect of mutagens and carcinogens is to identify the anticlastogenic, anti mutagenic (substances which suppress or inhibit the process of mutagenesis by acting directly on the mechanism of cell), and desmutagens (substances which somehow. destroy or inactivate, partially or fully the mutagens, thereby affecting less cell population) in our diets and trying to increase their consumption.

1.6.4 Replicative index assay (RI)

The examination of the cell cycle has proven to be an accurate and effective method for assessing cell health by analyzing cell proliferation in great details, also recognized as the "Circle of Life".

The cell cycle can be divided into two distinct stages :the first stage is inter phase which consists of the G1, S, and G2 phases where cells are active, growing, and the DNA replication takes place. The second stage is M-phase, also known as the "mitotic phase", where cell division takes place (Adams,2005). which consists of three phases (M1,M2,M3). The progression of cell division can be measured by culturing the case in the presence of bromodeoxyuridin (Brd U) after mitogenic stimulation and staining the fixed preparation using technique for obtaining differentially-stained chromosomes (Goto *et al* .,1978). In this test the cells that have replicated their DNA once in the presence of BdrU, will contain evenly and lightly –stained chromatids (M1). Those that have replicated twice in the presence of BdrU will contain differentially-stained chromatids, one light blue and other is dark blue (M2), when Giemsa stain is used. Cell that progress through three or more cell cycles in the presence of BdrU will contain differentially-stained and some evenly and lightly –stained chromosomes (M3) (Tan ., 1995; Perry and Wolff,1974). So Cell progression reveal the ability of some chemical and physical agents to inhibit the cell proliferation and replication . When the mitotic cells fail to reach the third mitotic cycle (M1). Many food compounds showed good results in inhibition of cell cycle . Reseratrol is one of these compounds that found in grapes that inhibits that growth and made the cell cycle arrest in G2/M phase (Holland et al .,2002). So the replicative index assay is aparameter that defined cell cycle progression, and considers the number of M1,M2 and M3 metaphases per 100 cells (Eke&. Çelik., 2008) .

2-1 Materials and methods

2.1.1. Equipments and Apparatus

Table (2-1): Instruments used in this study and their manufacturers.

No.	Apparatus	Origin	Company
1	CO_2 incubator	England	Gallenkamp
2	Centrifuge	Germany	Universal 16A
3	Balance	Switzerland	Mettler
4	Water Bath	England	Gallenkamp
5	Light Microscop	Japan	Olympus
6	Vortex	Germany	Buchi
7	Distiller	England	Exelo
8	Sensitive Electronic Balance	Switzerland	Mettler
9	pH-meter	England	Kent Industrial Ltd.
10	Oven	Germany	Memmert
11	Refrigerator	Japan	Sony
12	Nalgen Filter (0.22Mm)	Germany	Sartorius
13	Hot Plat with Magnetic Stirrer	England	Gallenkamp
14	Laminar air flow hood	England	Gelman Instrument U.K
15	Rotary evaporator	Yamato	Japan

2.1.2. Chemical Material

Table (2-2): Chemicals, used in this study and their supplier

No.	Chemical materials	Origin	Manufacturer company
1	Phytohemagglutinin (PHA)	Iraq	The Iraqi Center for Cancer Research and Medical Genetics(ICRM)
2	Colcimid	Iraq	(ICRM)
3	Cytochalacin –B (Cyt-B)	USA	Sigma
4	(RPMI-1640) Rosswell Park Memorial Institute -1640	USA	Sigma
5	Heparin	Denmark	Pharmaceutical
6	Bovine Serum	Iraqi	(ICRM)
7	Acetone	England	BDH
8	Gimsa stain	U.K	BDH
9	Potassium Chloride	England	BDH
10	Sodium hydroxide(NaOH)	U.K	BDH
11	Sodium bicarbonate(NaHCO3)	England	BDH
12	Glacial Acetic Acid (%99.9)	England	BDH
13	cupric sulfate ($CuSO_4$)	England	BDH
14	$(CH3COO)_3Pb$	England	BDH
15	Ferric chloride ($FeCl_3$)	England	BDH
16	Sodium chloride (NaCl)	England	BDH
17	Hydrochloric acid (HCl)	England	BDH
18	Chromic acid	Germany	Prolabo
19	Chloroform ($C_6H_{12}O_6$)	England	BDH
20	Dipotassium hydrogen phosphate(K2HPO4)	England	BDH
21	Potassium dihydrogen phosphate (KH2PO4)	England	BDH
22	Na2HPO4	England	BDH
23	Na2H2PO4	England	BDH
24	(%95) Ethanol	England	BDH
25	(%70) Ethanol	England	BDH
26	Methanol (%99.8)	England	BDH
27	Dimethyle Sulphoxide (DMSO)	England	BDH

2.2. Methods
2.2.1. Preparation process of pomegranate molasses.

We used fresh pomegranate juice (PJ) made from a pomegranate variety grown in Iraq. Pomegranate molasses is produced by peeling the fruits, dispersing the grains and pressing them manually to have a juice. The juice is boiled for more than six hours in order to obtain a concentrated substance called "molasses.". (Poyrazoglu et al., 2002; Incedayi et al., 2010). Attended concentrations (5, 10, and 15) µg /ml of pomegranate molasses where prepared.

Fig(2- 1) pomegranate molasses

2.2.2 Benzoic acid preparation

Benzoic acid (Sigma) was dissolved in 100 ml distilled water to make the concentration of 500µg / ml (the amount used in foods) (Yilmaz et al., 2009).

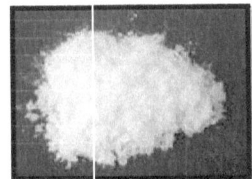

Fig(2- 2) Benzoic acid crystals

2.2.3 Preparation for cytogenetic study on human blood lymphocyte.

The following solution should be prepared (Yaseen et al., 1998):

2.2.3.1 Antibiotic solutions
Penicillin and streptomycin were used. The penicillin (1000000 IU) was dissolved in 10/ml of sterilized distilled water, while one gram of streptomycin was dissolved in 10/ml of sterilized distilled water. Both solutions were sterilized through Millipore filter (0.22µm) dividing them into aliquots each contains 1ml in a sterile vial (Freshney, 2000) divided to 1ml in sterile vial and stored at -20°C until use.

2.2.3.2 Growth medium: RPMI-1640 medium was prepared as follows:

RPMI media powder with glutamine	10.4 gm
Heps powder	2.5 gm
Sodium bicarbonate	15 ml
Penicillin	1.5 ml
Streptomycin	0.5 ml
Bovine serum	200 ml

The volume was completed with distilled water to get 1 litter, sterilized by millipore sterile filter 0.22 µm, at pH 6.8-7.2, then stored at –20°C

2.2.3.3 Colchicines solution :- was prepared by dissolving 1.0 mg of colchicines in 10 ml D.W., filtered, and stored at 4°C. (Allen *et al.*, 1977).

2.2.3.4 Hypotonic potassium chloride solution (0.075 M) : was done by dissolving 1.1175 g of KCl powder in 200 ml D. W., and was stored at 4°C.

2.2.3.5 The fixative:- Freshly made mixture of methanol and glacial acetic acid in the ratio 3:1 (v/v) (Allen *et al.*, 1977).

2.2.3.6 Sorenson's buffer :- It was prepared by dissolving 7.08g of Na_2HPO_4 with 7.74g of KH_2PO_4 in 1L of distilled water at pH 7 and stored in a refrigerator until use (Al-Rikabi, 1997).

2.2.3.7 Giemsa stain in which a stock solution was prepared by dissolving 2g of Giemsa stain powder in a 100 ml of methanol, stirred using magnetic stiffer at room temperature for two hours, filtered by filter paper (Wattman No.1) and stored in a dark tight bottle. Where staining, 1 ml of the stock solution is added to 4 ml of Sorenson's buffer (Sharma and Sharma, 1980).

2.2.3.8 Phytohaemagglutinin (PHA): It was supplied as a liquid by ICCMGR.

2.2.4 The Preparation of Human Chromosome aberrations from Peripheral Blood (Galloway, 2000).

2.2.4.1 Blood collection : Blood samples were collected from 10 healthy donors (five males and five females, non-smokers, with age range 25-28), by vein puncturing using a disposable syringe, 5 ml of blood was transferred into heparinized steril tubes.

2.2.4.2 Blood culturing

A. Four groups test of tubes where prepared as will be mentioned in section (2.4.3).

B. Half ml of peripheral blood was added to test tubes containing 5 ml of culture media (prepared in section 2.2.3.2)

C. Phytohaemagglutinin (0.3 ml) was added to all test tubes, then placed in the incubator at 37°C for 24 h.

D. All tubes were returned to the CO_2-incubator for 70 hours, and gently shaken each 12 hrs.

E. After 70 hrs of incubation, 0.1 ml of colchicines solution (prepared in section 2.2.3.3) was added to each test tube and complete incubation to 72 hrs.

F. Samples were centrifuged for 10 minutes at 1500 rpm.

G. Supernatant was withdrawn and about 0.5ml of supernatant was left over, and mixed well, then 5-10 ml of warmed Potassium chloride (0.075 M) was added with mixing.

H. Samples were incubated in the shaker water bath for 20 minutes at 37°C then centrifuged for 10 minutes at 1500 rpm, and then the supernatant discarded.

I. Few drops of freshly made fixative (methanol and glacial acetic acid 3:1(v/v)) were added to the tubes with gentle mixing until reaching 5ml, centrifuged for 10 minutes at 1500 rpm then fixative decanted off. The process repeated for 2-3 times.

J. Finaly, the cells re-suspended in a 3 ml of freshly made fixative and stored at (–20°C).

2.2.5 Cytogenetic experiments

This experiment was carried out to assess the protective effect of pomegranate molasses extract at three different concentrations (5, 10 or 15) µg/ml, against genotoxicity induced by benzoic acid (E-210) at concentration 500µg/ml, using three types of transactions (before, after, and with treatment) in blood human lymphocytes in vitro.

- Group 1: Negative control (only blood without treatment).
- Group 2: Positive control (blood treated with (500µg/ml) of benzoic acid).
- Groups 3 : Pre- treatment with Benzoic acid (E-210). The human lymphocytes culture were first treated with three concentrations (5, 10 and 15 µg/ml respectively with pomegranate molasses extract. After 24 hrs of incubation benzoic acid 500 µg/ml was added , complete incubated , to 70 hrs
- Groups 4 : Post- treatment with Benzoic acid (E-210). blood was first treated with (500µg/ml) of benzoic acid, than after 24 hrs of incubation, pomegranate molasses for was added with three doses (5, 10 and 15 µg/ml) respectively, and then complete incubated , to 70 hrs.

- Groups 5 : Simultaneous Treatment with Mixture of pomegranate molasses and Benzoic acid (E-210). Blood was treated with (500µg/ml) of benzoic acid and three doses (5, 10 or 15 µg/ml respectively of pomegranate molasses extract, The cultures were incubated at 37°C for 70 hours.

After 70 h of incubation, 0.1 mL of colcemid solution (1 µg/ml) was added for each groups , and then complete incubated, to 72 hrs.

2.2.6 Slide preparation

The procedure was followed according to ICCMGR protocol. The cell suspension was removed from freezer, thawed and centrifuged at 1500 rpm for 10 minutes. The supernatant was decanted off and the cells re-suspended in appropriate amount (0.5 ml) to make thinly cloudy suspension than 3-4 drops of cells suspension were dropped evenly from appropriate distance 30 cm onto a wet, chilled, grease-free slides and allowed to dry at room temperature. Four slides were prepared for each cytogenetic assay (M I, CA, MN, and RI).

2.2.7 Staining

The slides were stained using freshly made stain [Giemsa stain stock solution and Sorenson's buffer 1:4 v/v], washed with Sorenson's buffer . Microscopic examination under low magnification using 100X were made.

2.2.8 Cytogenetic parameter Analysis scoring

- **Mitotic index (MI) analysis**: The MI was determined as a ratio of the mitotic cells to the cells in interphase in 1000 calculated cells.

    M.I. % = (No. of Dividing cells / No. of dividing cells + No. of non-dividing cells) x 100 (Jordan *et al.*, 1996).

- **Chromosomal Aberrations (CAs) Assay :** The prepared slides were examined under the oil immersion lens of light microscope for 100 divided cells per blood lymphocytes culture, and the cells should be at the first metaphase stage of the mitotic division where the chromosomal aberrations are clear and the percentage of these aberrations was estimated (Hagmar *et al.*, 1998; Honma,2011)
- **Micronuclei Assay (MN):** In order to detect the number of micronucleated lymphocytes, cytochalasin B (4.5 µg/ml, Sigma) was added to cultures at 44th hour of incubation (Lorge *et al.,* 2006). At the end of the 72 h scoring The criteria for micronuclei were as described by Fenech *et al* ,(2003). At least 2000 binucleated lymphocytes were examined for the presence of one , two or more micronuclei per concentration (Ozkul *et al*, 2005).
- **Replicative Index (RI):** was determined by counting the number of cells at the first, second and the third metaphase in (100) a cell at metaphase, the RI was calculated according to the following equation:

    RI= (1xM1%) + (2xM2%) + (3xM3%)/100 (Eke&. Çelik., 2008) .

2.2.9 Statistical analysis

The SAS (2010) program was used to study the effects of treatments in different trails. The least significant difference (LSD) test was used to signify a comparison between the means.

3. Results and Discussion

3.1 Cytogenetic effects of pomegranate molasses (PM) extracts and benzoic acid (E210) in vitro :

3.1.1 Effects of pomegranate molasses (PM) and benzoic acid (E210) on mitotic index in lymphocytes culture .

The results as shown in Table (3-1) demonstrated that benzoic acid at 500 μg / ml resulted in significantly decrease of mitotic index in human lymphocytes compared with untreated control. It was 2.25 in comparison with the negative group(6.55). Whereas no significant differences ($p < 0.05$) in mitotic index resulted when the human lymphocytes treated with three concentration (5,10 and 15μg/ml) of pomegranate molasses , the mitotic index was 3.46 and 4.50, 5.65 respectively in comparison with the negative group.

Table (3-1) Cytogenetic effect of pomegranate molasses (PM) extract and benzoic acid (E210) on mitotic index in human blood lymphocyte culture (in vitro)

Test substance	Concentration μg /mL	Mitotic index
Control (negative group)	0	6.55 a
Benzoic acid	500	2.25 b
pomegranate molasses (PM)	5	3.46 c
	10	4.50 d
	15	5.65 e

Differences A, B, C, D, E are significant ($P < 0.05$) to compression rows

3.1.2 Effects of pomegranate molasses (PM) extract and benzoic acid (E210) on chromosomal aberrations in lymphocytes culture.

Cytogenetic data obtained with various concentrations of pomegranate molasses (5,10 and 15), on cultured blood cells, respectively as shown in table (1-2). The pomegranate molasses at tested concentrations did not induce significanly ($p < 0,05$) on frequencies of CAs when compared with negative control. However, the lymphocytes culture treated with benzoic acid (E210) (positive control) showed about three fold increases on total chromosomal aberrations parameters that were studied (chromatid and chromosome breaks, chromatid and chromosome gap, fragments, and dicentric) as compared with negative control.

Table (3-2). Effect of pomegranate molasses (PM) extract and benzoic acid (E210) on chromosomal aberrations in lymphocytes culture.

Test substances	Concentration (µg/ml)	Structural chromosomal aberration						Total %
		Acentric	Dicenteric	Gap chromosome	Gap chromatid	Break chromatid	Break chromosome	
Negative control	0	0.0 a	0.0 a	0.020 a	0.244 a	0.0 a	0.0 a	0.264 a
Benzoic acid (E210)	500	0.050 b	0.04 b	0.016 b	2.89 b	0.017 b	0.014 b	7.025 b
Pomegranate molasses (PM)	5	0.0 a	0.0 a	0.180 a	0.290 a	0.0 a	0.0 a	0.570 a
	10	0.0 a	0.0 a	0.012 a	0.420 a	0.0 a	0.0 a	0.432 a
	15	0.0 a	0.0 a	0.018 a	0.250 a	0.0 a	0.0 a	0.268 a

Differences A, B, C, D, E are significant (P< 0.05) to compression rows

3.1.3 Effects of pomegranate molasses (PM) extract and benzoic acid (E210) on Micronucleus (MN) formation in lymphocytes culture.

Benzoic acid significantly increased micronucleus frequency in lymphocytes ($P \leq 0.05$) as compared with negative control as show with table (3-3). while, we found that lymphocytes treatment with three concentrations (5,10 and 15µg/ml) of pomegranate molasses (PM) showed no significant difference ($p < 0.05$) in MN frequency compared with negative control groups.

Table (3-3) Effect of pomegranate molasses (PM) extract and benzoic acid (E210) on Micronucleus (MN) formation in lymphocytes culture

Test substance	Concentration µg/mL	Distribution of MN in BN				MN%
		0MN	1MN	2MN	3MN	
Negative control	0	998	1	0	6	0.7a
Benzoic acid (E210)	500	959	18	10	5	3.3 b
Pomegranate molasses (PM)	5	993	13	2	0	1.5a
	10	996	12	1	0	1.3a
	15	994	10	0	0	1a

Differences A, B, C, D, E are significant ($P < 0.05$) to compression rows

3.1.4 Effects of pomegranate molasses (PM) extract and benzoic acid (E210) on replication index (RI) in lymphocytes culture.

In this study, benzoic acid causes decrease in the replication index (RI), compared with negative control, mean while all the tested concentrations of pomegranate molasses (PM) did not lead to a marked decrease in the RI in comparison with the control group, as shown in table (3-4).

Table (3-4) Effect of pomegranate molasses (PM) extract and benzoic acid (E210) on replication index (RI) in lymphocytes culture

Test substance	Concentration	Cell cycl Progression %			(RI) %
		1M	2M	3M	
Negative control	0	34	28	40	2.26a
Benzoic acid (E210)	500	72	117	11	1.69b
pomegranate molasses (PM)	5	80	69	51	1.86b
	10	50	76	74	2.12a
	15	40	83	77	2.19a

Differences A, B, C, D, E are significant (P< 0.05) to compression rows
RI: Replication Index
SD: Standard Deviation
M1: The number of cells in first metaphase
M2: The number of cells in second metaphase
M3: The number of cells in third or more metapzz
RI = (1xM1 + 2xM2 + 3xM3)/500

In vitro genotoxicity tests detect compounds that induce genetic damage directly or indirectly by various mechanisms. One of these test systems is CA which has been considered as an early warning signal for cancer development (Hagmar et al.1998; Bonassi et al. 2000). These changes presumably involve DNA breakage and reunion (Pandita ,1988). Latt and Schreck (1981) proved SCEs to be a highly sensitive indicator for assessing potential mutagens and carcinogens. MN assay detects both clastogenicity (chromosome breakage) and aneugenicity (chromosome lagging due to dysfunction of mitotic apparatus) (Albertini et al. 2000).

The results obtained by us indicate that the treatment of blood lymphocytes with different concentrations of pomegranate molasses (PM) at all concentrations, were almost similar to the controls. which confirms that the pomegranate molasses does not possess the effects of genetic toxicity or mutagenic on blood lymph. This could be attributed to the pomegranate and its products are rich in phytochemicals and antioxidant. The antioxidant activity was almost determined via in vitro trials and several methods , however, pomegranate showed antioxidant activity, independent on the antioxidant test assayed and generally with significant linear correlation between phenolic content and antioxidant capacity (Naczk and Shahidi , 2004; Stratil et al., 2006; Moon and Shibamoto 2009; Deng et al., 2011). Chalfoun- Mounayar et al.(2013)

found that the pomegranate molasses and juice possesses a powerful antioxidant activity.

In our study, benzoic acid is able to significantly increased the chromosomal aberration, and micronucleus frequency in human lymphocytes, and cause a decreased in the mitotic and replication index. Benzoic acid induced six types of structural aberrations, the most common aberrations are chromatid breaks and gap chromatid union (Murli, 2003). There are many studies that showed the genotoxicity of different food additives in different cell lines (Macioszek and Kononowicz 2004; Sarıkaya and Cakır 2005; Yılmaz *et al.* 2008; Mpountoukas *et al.* 2008).

3.2 Interaction between benzoic acid and pomegranate molasses extract on blood lymph

This experiment was designed to know the effect of pomegranate molasses extract on mutagenic effect of benzoic acid which shows a high percentage of CA , increase in MN and decrease in MI and RI on blood lymph . Therefore; we would like to reveal if pomegranate molasses extract at three concentrations has the ability to reduce the effect of benzoic acid in 500µg/ml concentration by three treatment step (before, after and with benzoic acid).

3.2.1 Pre –benzoic acid treatment with pomegranate molasses extract:

The results of the pre -treatment with pomegranate molasses (PM) are shown in Table (3-6) and (3-7) significant decrease (P <0.05) of three concentrations of the pomegranate molasses (PM) extract on (CA, MN) parameters that were studied compared with treatment with benzoic acid (E210) of concentration 500 µg/ml resulted in increase in the frequencies of MN and CA alone. Whereas, The results of the pre - treatment with pomegranate molasses (PM) A significant increase in the frequencies of RI and MI parameters that were studied, compared with treatment with benzoic acid (E210) of concentration 500µg/ml induced resulted in a decrease in (RI, MI), there was no significant difference between these groups and negative control in most of the total parameters which was analyzed. It was clear from the Table (3-8) and (3-5) that all three concentration of pomegranate molasses (PM) extract were minimized the effect of benzoic acid (E210) in respect with total parameters, but the highest protective concentration was the third concentration (15 µg/ml) and the lowest effective concentration was the first concentration (5 µg/ml).

3.2.2 Post-benzoic acid treatment with pomegranate molasses extract

The results of post–treatment of blood lymph with three concentration of pomegranate molasses (5,10, and 15 µg/ml) respectively after the treatment with benzoic acid, did not eliminate the mutagenic effect of benzoic acid completely as in the negative control group, but it set it back to insignificant levels of benzoic acid effect in all parameters that studied, compared with the positive control (treated blood lymph with the benzoic acid only) Whereas, the concentration 15µg/ml scored the best results in the evaluation of all parameters that were studied (MI, MN, CA, RI)that formed a significant difference at the level of the probability $P <0.05$ when compared with the positive control, tables (3-5),(3-6),(3-7) and (3-8).

3.2.3 Simultaneous treatment benzoic acid with pomegranate molasses extract

The results of simultaneous treatment benzoic acid with pomegranate molasses extract showed in tables (3-5),(3-8). The results indicated that the positive control significant decrease in mitotic index and RI as compared with negative control, while, showed highly increase in chromosomal aberrations and MN on blood lymph as compared with negative control tables (3-6),(3-7). While, the results of co-treatments, of blood lymph with three concentration (5,10,15 µg/ml) of pomegranate molasses extract, and benzoic acid (500µg/ml), was significantly decrease the RI and MI, table (3.6) and (3-7) a significant reduction in CAs , MN formation, compared with positive control (the treated blood lymph with the benzoic acid only).

Table 3.5 Interaction between pomegranate molasses (PM) extract and benzoic acid (E-210) on mitotic index in human blood lymphocyte culture (in vitro)

Test substance	Concentration µg /mL	Mitotic index
Control	0	6.55 a
Benzoic acid	500	2.25 b
Post - benzoic acid treatment	5	3.55 c
	10	3.72 c
	15	4.35 d
Pre- benzoic acid treatment	5	5.55 a
	10	5.85 a
	15	6.34 a
Simultaneous treatment	5	5.35 a
	10	5.54 a
	15	6.23 a

Differences A, B, C, D, E are significant (P< 0.05) to compression rows

Table 3.6 The effect of interaction between pomegranate molasses extract and benzoic acid (E210) on chromosomal aberrations (CA) in human blood lymphocyte culture (in vitro)

Test substance	Concentration µg/mL	Chromosomal Aberrations (CA)					Total %
		Dicenteric	Chromatid break	Chromosome break	Gap	Acentric	
Control	0	0	0	0	2	0	0.02 a
Benzoic acid	500	0.34	1.52	0.35	2.27	0.03	4.52 b
Post- benzoic acid treatment	5	0.14	0.85	0	0.98	0	1.97 c
	10	0	0.54	0	0.76	0	1.3 c
	15	0	0.10	0	0.56	0	0.66 a
Pre - benzoic acid treatment	5	0	0.432	0	0.32	0	0.752 a
	10	0	0.05	0	0.09	0	0.14 a
	15	0	0.02	0	0.04	0	0.06 a
Simultaneous treatment	5	0	0.64	0	0.55	0	1.19 c
	10	0	0.01	0	0.2	0	0.21 a
	15	0	0.05	0	0.01	0	0.06 a

Differences A, B, C, D, E are significant (P< 0.05) to compression row

Table 3.7 The effect of pomegranate molasses extract on benzoic acid induced micronuclei formation in human blood lymphocyte culture (in vitro)

Test substance	Concentration µg /mL	No. of cells carrying micronuclei				MN %
		0MN	1MN	2MN	3MN	
Control	0	995	5	0	0	0.5a
Benzoic acid	500	965	19	5	2	3.5b
Post- benzoic acid treatment	5	978	14	5	1	2.7c
	10	986	15	3	0	2.1c
	15	988	10	4	0	1.8c
Pre- benzoic acid treatment	5	990	13	2	0	1.7c
	10	995	10	1	0	1.2c
	15	996	9	0	0	0.9a
Simultaneous treatment	5	988	10	1	0	1.1c
	10	989	8	1	0	1.0c
	15	993	7	0	0	0.7a

Differences A, B, C, D, E are significant ($P < 0.05$) to compression row
MN: Micronucleus

Table 3.8 The effect of pomegranate molasses extract ((PM)) on replication index (RI) in human lymphocytes induced by benzoic acid

Test substance	Concentration µg /ml	Cell cycle Progression %			RI%
		1M	2M	3M	
Control	0	30	60	14	1.92a
Benzoic acid	500	38	55	7	1.69b
Post- benzoic acid treatment	5	36	57	7	1.71b
	10	30	44	20	1.78a
	15	20	60	12	1.76b
Pre- benzoic acid treatment	5	35	62	7	1.80a
	10	28	55	15	1.83a
	15	25	63	12	1.87a
Simultaneous treatment	5	33	55	14	1.85a
	10	32	60	12	1.88a
	15	33	56	15	1.90a

Differences A, B, C, D, E are significant ($P < 0.05$) to compression
RI: Replication Index
RI = (1×M1 + 2×M2 + 3×M3)/500
M1: The number of cells in first metaphase
M2: The number of cells in second metaphase
M3: The number of cells in third or more metapzz

The results obtained in this study in the pre-treatment with benzoic acid could reflect the effects of pomegranate molasses on reducing genotoxicity of benzoic acid, which could be associated to the antioxidant acting or interfering on DNA replication. The results observed in the simultaneous treatment suggest that pomegranate molasses concentrations exhibit des mutagenic potential, protecting chromosomes by direct action on the mutagen and inactivating it, therefore reduce the effect of these mutagenic materials and their metabolites(Kanakis et al.,2005). It is possible to consider pomegranate molasses extract as a desmutagens for its ability to decrease the effect of benzoic acid by chemical inactivates, enzymatic inducers, mutagen scavenger, or as antioxidants.

In the post-treatment, no significant results were observed, pointing out that pomegranate molasses does not play any role in DNA repair mechanisms. This indicate that pomegranate molasses exert high protective effect in case of pre- treatment or simultaneous treatment, therefore, considered as desmutagen activity more than biomutagen activity (Bronzetii, 1997).

It is clear that bioactive compounds present in daily diet, mainly in fruits and vegetables, have a prevention potential in cancer by inhibiting carcinogenesis through cell-defensive and cell-death mechanism regulation. These chemopreventive effects may be attributed to a complex effect of various phenolic substances of antioxidant capacity (Khan et al., 2009). Pomegranate is rich in anthocyanins, 3-glucosides, 3,5-diglucosides of delphinidin, cyanidin and pelargonidin, ellagitannins and other phenolic compounds, which are known bioactive compounds with antioxidant and antitumoral activity (Chaturvedula et al., 2011; Zhang et al., 2011). Proestos et al. (2013) reported that dietary polyphenols are currently attracting much interest because of their antioxidant, anti-inflammatory, and anticarcinogenic effects in order to establish conclusive evidence for the effectiveness of polyphenols in disease prevention and human health improvement, it is essential to determine the distribution of these compounds in our diet, estimating their content in each food, and to identify which of the hundreds of existing polyphenols are likely to provide the greatest effects in the context of preventive nutrition (Porrini et al .,2008).

Pomegranate molasses has been described as an important source of natural phenolic antioxidants such as (tannins, ellagic and gallic acids) (Jurenka, 2008). In fact, the phytochemical analysis of pomegranate molasses identified the presence of flavonoids, phenolic compounds and saponins (Fadavi et al ., 2005; Yilmaz et al.,2007). In this context, the protective activity of pomegranate molasses should be associated to its constitutive antioxidant compounds. according Chalfoun-Mounayar et al. (2012) the polyphenols in molasses are four times greater than those found in the juice, moreover, it has the strongest antioxidant properties in vitro compared to pomegranate juice.

Incedayi et al ., (2010) reported that pomegranate molasses are also rich in phenolics with high health and nutritional values, if it is produced in the correct way, therefore, to improve the production techniques of pomegranate molasses it is very necessary to limit the hydroxymethyl furfural (HMF) content, which was the result of long term thermal processing. The hydroxymethyl furfural (HMF) content indicates the degree of heating of the treated products during processing and the quantification of this compound is considered as a quality parameter for concentrated food products with adverse side effects (Ramirez-Jimenez et al., 2000; Rada-Mendoza et al., 2002; Kus et al., 2005).The temperature at limits of the standard does not alter the antioxidant activity of pomegranate molasses against reactive oxygen species (ROS) (Anonymous ,2001), On the other hand that the standard temperature helps polyphenols to be released from pomegranate fruit cell as there is no extraction with a solvent in the preparation of pomegranate molasses. (Yılmaz et al., 2007).

Conclusions

It can be concluded from this study that benzoic acid is genotoxic agent especially in 500 µg/ml concentration used in human lymphocyte cultures. our results confirmed the safe use of pomegranate molasses in the tested concentrations, hence that the results indicated that the pomegranate molasses extract at concentrations applied in the present study a Protective effect by modulating the cytogenetic Effects caused by benzoic acid

Recommendations

- Further studies on the effect pomegranate molasses on cell lines in vitro studies are needed.
- Identification and characterization of the compounds present in the pomegranate molasses to determine their particular function will be part of the future studies.
- We propose that it is necessary to be careful when using industrial food that contain food additives such as preservatives especially benzoic acid. However, in relation to the genotoxic effect of benzoic acid, and for better understand, further genotoxicity studies are needed
- Having food that contain well prepared pomegranate molasses to a chief beneficial health , and clinical studies are needed to clarify some of its beneficial roles in health.

Reference

1. Abd Elmonem. H. A.(2014). Assessment the Effect of Pomegranate Molasses against Diazinon Toxicity in Male Rats . www.iosrjournals.org, 8(2): 135-141

2. Adams, D.J. (2005) Chapter 6 – Cell Cycle. University of Michigan Flow Core, USA

3. Adams TB, Cohen SM, Doull J, Feron VJ, Goodman JI, Marnett LJ, Munro IC, Portoghese PS, Smith RL, Waddell WJ, Wagner BM. 2005. The FEMA GRAS assessment of benzyl derivatives used as flavor ingredients. Food Chem Toxicol 43:1207–40

4. Adams, L.S., Seeram, N.P., Aggarwal, B.B., Takada, Y., Sand, D. & Heber, D. (2006). Pomegranate Juice, Total Pomegranate Ellagitannins, and Punicalagin Suppress Inflammatory Cellsignaling in Colon Cancer Cells. Journal of Agricultural and Food Chemistry,Vol.54, No.3, pp.980-985, ISSN 0021-8561

5. Adsule, R.N., Kotecha, P.M. & Kadam, S.S. (1992). Preparation of Wine from Pomegranate. Beverage Food World,Vol.19, No.4, pp.13

6. Adsule, R.N. & Patil, N.B. (1995). Pomegranate: InHandbook of Fruit Science and Technology, Salunke, D.K. & Kadam, S.S. (Eds), pp. 455-464, ISBN 0824796438, Marcel Dekkar, New York

7. Albertini, RJ.; Anderson, D.; Douglas, GR. (2000). IPCS guidelines for the monitoring of genotoxic effects of carcinogens in humans. Mutat ResRev., 463: 111-72.

8. Allen, J. W.; Shuler, C. F.; Menders, R. W. and Olatt, S. A. (1977). A simplified technique for in vivo analysis of sister chromatid exchange using 5-bromodeoxy uridine tablats. J. Cytogenet. Cell. Genet., 18: 231-237.

9. Al-Maiman, S.A. & Ahmad, D. (2002). Changes inPhysical and Chemical Properties during Pomegranate (Punica granatum L.) Fruit Maturation. Food Chemistry,Vol.76, No.4, pp.437-441, ISSN 0308-8146

10. AL-Rikabi, A.G. (1997). Cytogenetic and reproductive study of cobbelt effect on the working in industrial Battery in Iraq. Ph.D. thesis. College of science - AL-Mustensiria University.

11. Altug˘ T (2003) Introduction to toxicology and food, food additives. CRC Press, New York, pp 81–100

12. Alzahran, H.A.S. (2011) Protective effect ofL-carnitine against acrylamide-induced DNA damage in somatic and germ cells of mice . Saudi Journal of Biological Sciences 18,29–36

13. Anonymous, 2001. TSE 12720 Nar ekşisi standardı. Türk StandartlarıEnstitüsü. Necatibey Caddesi, 112, Bakanlıklar-Ankara.

14. Armstrong, M.J. and Galloway, S.M.(1994). Micronuclei induced in peripheral blood of mu-PIM-2 transgenic mice by chronic oral treatment with 2-Acetyl aminofluorene or benzene but not with diethylnitorosamine or 1, 2-Dichloroethane. Mutation Research, 302: 61 – 70.

15. Asif M. The role of fruits, vegetables, and spices in diabetes. Int J Nutr Pharmacol Neurol Dis 2011;1:27-35.

16. Ávila .R. I., M. T. Guerra, K. A. S. Borges, M. S. Vieira, L. M. Oliveira Júnior, H. Furtado, M. F. Mota, A. F. Arruda, M. C. Valadares.(2013). Punica granatumL. protects mice against hexavalent chromium-induced genotoxicity , Brazilian Journal of Pharmaceutical Sciences ,49(4): 690-697.

17. Aviram, M., Dornfeld, L., Rosenblat, M., Volkova, N., Kaplan, M., Coleman, R., Hayek, T ,. Presser, D. & Fuhrman, B. (2000).

Pomegranate Juice Consumption Reduces Oxidative Stress, Atherogenic Modifications to LDL and Platelet Aggregation : Studies in Humans and in Atherosclerotic Apolipoprotein E-deficient mice. The American Journal of Clinical Nutrition, Vol.71, No.5, pp.1062-1076, ISSN 0002-9165.

18. Beezhold BL, Johnston CS, Nochta KA. 2014. Sodium benzoate-rich beverage consumption is associated with increase reporting of ADHD symptoms in college students: a pilot investigation. J Attention Disor 18:236–41

19. Belsito D, Bickers D, Bruze M, Calow P, Dagli ML, Fryer AD, Greim H, Miyachi Y, Sipes IG. 2012. A toxicological and dermatological assessment of aryl alkyl alcohols when used as fragrance ingredients. Food Chem Toxicol 50:552–99.

20. Bezar HJ, Conner AJ. GM Foods: Evolution or Revolution?, New Zealand Dietetic Association Conference 2002, Palmerston North, 2002. Vol. 7. New Zealand Dietetic Association.

21. Bonaccorsi G, Perico A, Colzi A, Bavazzano P, Giusto MD, Lamberti I, Martino G, Lorini C. 2012. Benzene in soft drinks: a study in Florence (Italy). Ig Sanita Pubbl 68:523–32.

22. Bonassi,S.; Hagmar,L.; Stromberg,U.; Montagud,A.H.; Tinnerberg,H.; Forni,A.; Heikkila,P.; Wanders,S.; Wilhardt,P.; Hansteen,L.; Knudson,L.E.; NorppaH. (2000). Chromosomal aberrations in lymphocytes predict human cancer independently of exposure to carcinogens, Cancer Res., 60:1619–1625.

23. Borochov-Neori, H., Judeinstein, S., Tripler, E., Harari, M., Greenberg, A., Shomer, I. & Holland, D. (2009). Seasonal and Cultivar Variations in Antioxidant and Sensory Quality of Pomegranate (Punica granatum L.) Fruit. Journal of Food Composition and Analysis,Vol.22, No.3, pp.189-195, ISSN 0889-1575

24. Brul S, Coote P. 1999. Preservative agents in foods: mode of action and microbial resistance mechanisms. Intl J Food Microbiol 50:1–17.

25. Bronzetti, G.(1997). The role of antimutagenesis and carcinogenesis. J.Environ. Patho. Toxical. Oncol. 16: 259-269.

26. Cakir R, Cagri-Mehmetoglu. 2013. Sorbic and benzoic acid in non-preservative-added food products in Turkey. Food Addit Contam 6:47–54

27. Cammack R, Joannou CL, Cui X, Martinez CT, Maraj SR, Hughes MN. 1999. Nitrite and nitrosyl compounds on food preservation. Biochim Biophys Acta 1411:475–88.

28. Cantarelli MA, Pellerano RG, Marchevsky EJ, Camiˉna JM. 2009. Simultaneous determination of aspartame and acesulfame-K by molecular absorption spectrophotometry using multivariate calibration and validation by high- performance liquid chromatography. Food Chem 115:1128–32.

29. Carrano, A.V. (1986). Sister chromatide exchange induction and persistence in peripheral blood and spleen lymphocyte of mice. Environmental Mutagenesis, 8 (3):345-355.

30. Chauhan, L.K. Kumar, M. Paul, B.N. Goel, S.k. and Gupta, SK. (2007). Cytogenetic effects of commercial formulations of deltamethrin and/or isoproturon on human peripheral lymphocytes and mouse bone marrow cells. Environ. Mol. Mutagen. 48 (8):636-643.

31. Chalfoun-Mounayar.F, Nemr.R, Yared.P, Khairallah. S and Chahine R. (2012) Antioxidant and Weight Loss Effects of Pomegranate Molasses. *J APS* 02 (06); 45-50.

32. Chaturvedula, V., Prakash, S. & Prakash, I. (2011). Bioactive Chemical Constituents from Pomegranate (Punica granatum) Juice, Seed and Peel-A Review. International Journal of Research on Chemistry and Environment, Vol.1, No.1, pp.1-18, ISSN 0306-7319

33. Chen C, Lin W, Kuo C, Lu F. 2011. Role of redox signalling regulation in the propyl gallate-induced apoptosis of human leukemia cells. Food Chem Toxicol 49:494–50

34. Cressey P, Jones S. 2009. Levels of preservatives (sulphite, sorbate and benzoate) in New Zealand food and estimated dietary exposure. Food Addit Contam 26:604–13

35. Dahlawi, H., N. Jordan-Mahy, M. R. Clench, and C. L. Le Maitre. 2012. Bioactive actions of pomegranate fruit extracts on leukemia cell lines in vitro hold promise for new therapeutic agents for leukemia. Nutr. Cancer 64:100–110.

36. Dai, Z., Nair, V., Khan, M. & Ciolino, H.P. (2010). Pomegranate Extracts Inhibits the Proliferation and Viability of MMTV-Wnt-1 Mouse Mammary Cancer Stem Cells In vitro. Oncology Reports, Vol.24, No.4, pp.1087-1091, ISSN 1021-335X

37. Darbre PD, Aljarrah A, Miller WR, Coldham NG, Sauer MJ, Pope GS. 2004. Concentrations of parabens in human breast tumours. J Appl Toxicol 24:5–13

38. Dengate S, Ruben A. 2002. Controlled trial of cumulative behavioural effects of a common bread preservative. J Paediatr Child Health 38:373–6

39. Deng J, Cheng W, Yang G. 2011. A novel antioxidant activity index (AAU) for natural products using the DPPH assay. Food Chem 125:1430–5

40. Dumas, Y., Dadomo, M., Di Lucca, G. & Grolier,P. (2003). Effects of Environmental Factors and Agricultural Techniques on Antioxidant Content of Tomatoes. Journal of the Science of Food and Agriculture, Vol.83, No.5, pp.369-382, ISSN 0022-5142

41. Ekanem, A. M. and Osuji, J. O. (2006). Mitotic index studies on edible cocoyam. African J. Biotech., 5: 846-849.

42. Eke, D. and A. Çelik. 2008. Genotoxicity of thimerosal in cultured human lymphocytes with and without metabolic activation sister chromatid exchange analysis proliferation index and mitotic index. Toxicol In vitro., 22(4): 927-934

43. Elfalleh, W., Nasri, N., Marzougui, N., Thabti, I., M'Rabet, A., Yahya, Y., Lachiheb, B., Guasmi, F. & Ferchichi, A. (2009).

Physico-chemical Properties and DPPH-ABTS Scavenging Activity of Some Local Pomegranate (Punica granatum) Ecotypes. International Journal of FoodSciences and Nutrition, Vol.60, No.s2, pp.197-210, ISSN 0963-7486

44. Fadavi A., Barzegar M., Azizi, M.H. and Bayat, M. 2005. Physicochemical composition of ten pomegranate cultivars (Punica granatumL.) grown in Iran. Food Science and Technology International. 11(2):113-119

45. Fadavi, A., Barzegar, M. & Azizi, H.M. (2006).Determination of Fatty Acids and Total Lipid Content in Oilseed of 25 Pomegranates Varieties Grown in Iran. Journal of Food Composition and Analysis, Vol.19, No.6-7, pp. 676-680, ISSN 0889-1575

46. Fahad, A.Q. and Salim (2009). Mutagenic effects of sodium azide and its application in crop improvement. World Applied Sciences J. 6(12):1589-1601.

47. Fatma E. Agha ,Mahrousa S. Hassannane, Enayat A. Omara, Azza M. Hasan and Sayed A. El-Toumy. (2013). Protective Effect of *Punica granatum* Peel Extract Against Pentachlorophenol-Induced Oxidative Stress, Cytogenetic Toxicity and Hepatic Damage in Rats. *Aust. J. Basic & Appl. Sci.*, 7(2): 853-864

48. Fenech M, Chang WP, Kirsch-Volders M, Holland N, Bonassi S, Zeiger E. (2003). HUMN project: detailed description of the scoring criteria for the cytokinesis-block micronucleus assay using isolated human lymphocyte cultures. *Mutat. Res* .534 :65–75.

49. Fenech, M. (2007). Cytokinesis-block micronucleus cytome assay. Nat.Protoc.,2:1084-1104.

50. Fenech, M. (2010). The lymphocyte cytokinesis-block micronucleus cytome assay and its application in radiation biodosimetry. HealthPhys., 98:234-243.

51. Frazier RA, Inns EL, Dossi N, Ames JM, Nursten HE. 2000. Development of a capillary electrophoresis method for the simultaneous analysis of artificial sweeteners, preservatives and colours in soft drinks. J Chromatogr A 876:213–20.

52. Freshney, R.I. (2000). Culture of animal cell: A manual for Basic Technique. Wiley liss, A John Wiley and Sons, Inc. Publications, New York.

53. Galloway, S. M. (2000) Cytotoxicity and chromosome aberrations in vitro: experience in industry and the case for an upper limit on toxicity in the aberration assay. Environ. Mol. Mutagen., 35, 191–201

54. Garc´ıa-Gav´ın J, Parente J, Goossens A. 2012. Allergic contact dermatitis caused by sodium metabisulfite: a challenging allergen. A case series and literature review. Contact Dermatitis 67:260–9

55. Garcıa-Jim´ enez MC, Capitan-Vallvey LF. 2007. Simultaneous determination of antioxidants, preservatives and sweetener additives in food and cosmetics by flow-injection analysis coupled to a monolithic column. Anal Chim Acta 594:226–33

56. Gil, M., Tomas-Barberan, I., Hess-Pierce, F., Holcroft, B., D. M., & Kader, A. (2000). Antioxidant Activity of Pomegranate Juice and its Relationship with Phenolic Composition and Processing. Journal of Agricultural and Food Chemistry, Vol.48, No.10, pp. 4581-4589, ISSN 0021-8561

57. Goto,k. medea, S. Kano,Y&Sugeyama.T.(1978).Factors involved in differential giemsa staining of sister chromatids . chromosoma 66:351-359.

58. Guleray, A. and Lokman, A. (2005). Antagonistic effect of selenium against aflatoxin G1 toxicity induced chromosomal aberrations and metabolic activities of two crop plants. Bot. Bull. Acad. Sin., 46: 301-305.

59. Gultekin F, Doguc DK, Husamettin V, Taysi E. 2013. The effects of food and food additives on behaviors. Intl J Health Nutr 4:21–32

60. Guo C, Wei J, Yang J, Xu J, Pang W, Jiang Y. Pomegranate juice is potentially better than apple juice in improving antioxidant function in elderly subjects. Nutr Res 2008;28:72-7

61. Gultekin F, Doguc DK, Husamettin V, Taysi E. 2013. The effects of food and food additives on behaviors. Intl J Health Nutr 4:21–32

62. Greenwald P, Clifford CK, Milner JA. Diet and cancer prevention. European Journal of Cancer 2001; 37:948-965

63. Hagmar, L.; Bonassi, S.; Strömberg, U.et al. (1998). Chromosomal aberrations in lymphocytes predict human cancer. a report from the European Study Group on Cytogenetic Biomarkers and Health (ESCH). Cancer Res., 58: 4117-4121.

64. Hagmar L, Stromberg U, Bonassi S, Hansteen IL, Knudsen LE, Lindholm C, Norppa H (2004) Impact of types of lymphocyte chromosomal aberrations on human cancer risk: results from Nordic and Italian cohorts. Cancer Res 64: 2258–2263

65. Hartman, R.E., Shah, A. & Fagan, A.M. (2006). Pomegranate Juice Decreases Amyloid Load and Improves Behavior in a Mouse Model of Alzheimer's disease. Neurobiology of Disease,Vol.24, No.3, pp.506-515, ISSN 0969-9961.

66. Holland, N., P. Duramad, N. Rothman, L. Figgs, A. Blair, A. Hubbard and M. Smith. 2002. Micronucleus frequency and proliferation in human lymphocytes after exposure to herbicide 2,4-dichlorophenoxyacetic acid In vitroand In vivo. Mutat Res., 521: 165-178.

67. Holland, D.F. (2005). Chromosome analysis : banding patterns and structural aberrations. The University of Texas Southwestern Medical Center at Dallas.

68. Honma M. (2011), Cytotoxicity measurement in vitro chromosome aberration test and micronucleus test, Mutation Res., 724, 86-87.

69. Hord NG, Tang Y, Bryan NS. 2009. Food sources of nitrates and nitrites: the physiologic context for potential health benefits. Am J Clin Nutr 90:1–10.

70. Hord NG. 2011. Dietary nitrates, nitrites, and cardiovascular disease. Curr Atheroscler Rep 13:484–92.

71. Hut, H. M.; Lemstra, W. A.; Blaauw, E. H; Van, G. W.; Kampinga, H. H.; and Sibon, O. C. (2003). Centrosomes split in the presence of impaired DNA integrity during mitosis. Mol. Biol. Cell., 14: 1990-2000.

72. Incedayi, B., Tamer, E.C. & Copur, U. (2010). A Research on the Composition of Pomegranate Molasses. Journal of Agricultural Faculty of Uludag University, Vol.24, No.2, pp.37-47, ISSN 1301-3165

73. Ishidate M Jr, Sofuni T, Yoshikawa K, Hayashi M, Nohmi T, Sawada M, Matsuoka A (1984) Primary mutagenicityscreening of food additives currently used in Japan. Food Chem Toxicol 22:623–636

74. Jadeja, R.N., Thounaojam, M.C., Patel, D.K., Devkar, R.V., Ramachandran, A.V. (2010). Pomegranate (Punica granatum L.) Juice Supplementation Attenuates Isoproterenolinduced Cardiac Nerosis in Rats. Cardiovascular Toxicology, Vol. 10, No:3, pp. 174-180, ISSN: 1559-0259

75. Jordan, M. A.; Wendell, K.; Gardiner, S.; Derry, W. B.; Copp, H., and Wilson, L. (1996). Mitotic block induced in HeLa cells by low concentrations of paclitaxel (Taxol) results in abnormal mitotic exit and apoptotic cell death. Cancer Res., 56:816–825.

76. Jorge, I.; Gonzalez, B.; Amadeu, C.; Ricard, M.; and Jorge, Z. (2003). The mutagenic potential of the furyl ethylene derivative 2-Furyl-1nitroethene in the mouse bone morrow micronucleus test. Toxicological Science, 72:359-362.

77. Jurenka JS. Therapeutic applications of pomegranate (Punica granatum L.): a review. Altern Med Rev 2008 Jun;13(2):128-44.

78. Kaya, A. & Sozer, N. (2005). Rheological Behaviour of Sour Pomegranate Juice Concentrates (Punica granatum L.). International Journal of Food Science and Technology,Vol.40, No.2, pp.223-227, ISSN 1365-2621

79. Khan, S.A. (2009). The Role of Pomengranate (Punica granatum L.) in Colon Cancer. Pakistan Journal of Pharmaceutical Science,Vol.22, No.3, pp. 346-348, ISSN 1011-601X

80. Khan, N., Afaq, F. & Mukhtar, H. (2008).Cancer Chemoprevention through Dietary Antioxidants: Progress and Promise. Antioxidants Redox Signaling, Vol.10, No. 3, pp.475–510, ISSN 1523-0864

81. Kirsch-Volders, M., Sofuni, T., Aardema, M., Albertini, S., Eastmond, D., Fenech, M., Ishidate, M. Jr., Kirchner, S., Lorge, E., Morita, T., Norppa, H., Surralles, J., Vanhauwaert, A. and Wakata, A. (2004), Corrigendum to "Report from the in vitro micronucleus assay working group", Mutation Res.,564, 97-100.

82. Kryston, T. B., Georgiev, A. & Georgakilas, A.G. (2011). Role of Oxidative Stress and DNA Damage in Human Carcinogenesis. Mutation Research, Vol.711, No.1-2, pp.193-201, ISSN 0027-5107

83. Kus, S., F. Gogus and S. Eren. 2005. Hydroxymethyl furfural content of concentrated food products. International Journal of Food Properties, 8:2, 367-375.

84. Lansky EP, Newman RA (2007). Punica granatum (pomegranate) and its potential for prevention and treatment of inflammation and cancer. *J Ethnopharmacol* 19;109(2):177-206.

85. Langley P. Why a pomegranate? BMJ 2000;321:1153-4.

86. Latt SA, Schreck RR (1981) Sister chromatid exchange analysis. Am J Hum Genet 32:297–313

87. Lino CM, Pena A. 2010. Occurrence of caffeine, saccharin, benzoic acid and sorbic acid in soft drinks and nectars in Portugal and subsequent exposure assessment. Food Chem 121:503–8.

88. Liu RH (2004) Potential synergy of phytochemicals in cancer prevention: mechanism of action. J. Nutr. 134:3479S-3485S

89. Lorge, E., Thybaud, V., Aardema, M.J., Oliver, J., Wakata, A., Lorenzon G. and Marzin, D. (2006), SFTG International collaborative Study on in vitro micronucleus test. I. General

conditions and overall conclusions of the study, Mutation Res., 607, 13-36.

90. Macioszek VK, Kononowicz AK (2004) The evaluation of the genotoxicity of two commonly used food colors: quinoline yellow (E 104) and brilliant black BN (E 151).Cell Mol Biol Lett 9:107–122.

91. Maqsood S, Benjakul S, Shahidi F. 2013. Emerging role of phenolic compounds as natural additives in fish and fish products. Crit Rev Food Sci Nutr 53:162–79.

92. Maier E, Kurz K, Jenny M, Schennach H, Ueberall F, Fuchs D. 2010. Food preservatives sodium benzoate and propionic acid and colorant curcumin suppress Th1-type immune responsein vitro. Food Chem Toxicol 48:1950–6

93. Mamur S, Yüzbasıoğlu D, Ünal F, Yılmaz S (2010) Does the potassium sorbate induce genotoxic or mutagenic effects in lymphocytes? Toxicol In Vitro 24:790–794

94. Martin, O.A., Redon, C., Nakamura, A J., Dickey, J.S., Georgakilas, A.G. & Bonner, W.M. (2011). Systemic DNA Damage Related to Cancer. Cancer Research, Vol.71, No.10, pp.1-5, ISSN 0304-3835

95. Martyn DM, McNulty BA, Nugent AP, Gibney MJ. 2013. Food additives and preschool children. Proc Nutr Soc 72:109–16.

96. Maskan, A., Kaya, S. & Maskan, M. (2002). Effect of Concentration and Drying Hot-Air and Microwave Drying. Journal of Food Engineering, Vol. 48, No.2, pp.169-175, ISSN 0260-8774

97. McCann D, Barrett A, Cooper A, Crumpler D, Dalen L, Grimshaw K, Kitchin E, Lok K, Porteous L, Prince E, Sonuga-Barke E, Stevenson J. 2007. Food additives and hyperactive behaviour in 3-yar-old and 8/9-year-old children in the community: a randomised, double-blind, placebo-controlled trial. Lancet 370:1560–7.

98. Meera, S. and Nagarjuna, G. C. (2010). Antimutagenic activity of aqueous extract of Momordica charantia.International J. for Biotechnology and Molecular Biology Research. 1(4):42-46.

99. Merusi C, Corradini C, Cavazza A, Borromei C, Salvadeo P. 2010. Determination of nitrates, nitrites and oxalates in food products by capillary electrophoresis with pH-dependent electro-osmotic flow reversal. Food Chem 120:615–20.

100. Mollinedo, F. and Gajabe, C. (2003). Microtubules, microtubuleinterfering agents and apoptosis. Apoptosis, 8: 413-450.

101. Moneim, A.E.A.; EL-Khadragy, M.F (2012).The potential effects of pomegranate (Punica granatum) juice on carbon tetrachloride-induced nephrotoxicity in rats. J. Physiol. Biochem., v.69, n.3, p.359-370.

102. Moon J-K, Shibamoto T. 2009. Antioxidant assays for plant and food components. J Agric Food Chem 57:1655–66.

103. Murli H (2003) Screening assay for chromosomal aberrations in chinese hamster ovary (CHO) cells with argentyn. Final report. Covance Lab 23:1–15

104. Mpountoukas P, Vantarakis A, Sivridis E, Lialiaris T (2008) Cytogenetic study in cultured human lymphocytes treated with three commonly used preservatives. Food Chem Toxicol 46:2390–2393.

105. Naczk M, Shahidi F. 2004. Extraction and analysis of phenolics in food. J Chromatogr A 1054:95–111

106. Natheer J. Yaseen, Mustafa S. Mustafa Al-Attar,(2014). Assessment of mutagenic and antimutagenic effects of Punica granatumagainst ifosfamide induced chromosomal aberrations in male albino mice, Iraqi Journal of Cancer and Medical Genetics,7(1):5-10.

107. Ohtsuki T, Sato K, Sugimoto N, Akiyama H, Kawamura Y. 2012. Absolute quantitative analysis for sorbic acid in processed foods using proton nuclear magnetic resonance spectroscopy. Anal Chim Acta 734:54–61.

108. Ozkul, Y., S. Silici and E. Eroğlu. 2005. The anticarcinogenic effect of propolis in human lymphocytes culture. Phytomedicine, 12: 742-747.

109. Paller, C. J., X. Ye, P. J. Wozniak, B. K. Gillespie, P. R. Sieber, R. H. Greengold, et al. 2012. A randomized phase II study of pomegranate extract for men with rising PSA following initial therapy for localized prostate cancer. Prostate Cancer Prostatic Dis. doi: 10.1038/pcan2012.20 [Epub ahead of print]

110. Pande, G. & Akoh, C.C. (2009). Antioxidant Capacity and Lipid Characterization of Six Georgia- Grown Pomegranate Cultivars. Journal of Agricultural and Food Chemistry, Vol.57, No.20, pp.9427-9436, ISSN 0021-8561

111. Park, H.M.., Moon, E., Kim, A.J., Kim, M.H., Lee, S., Lee, J.B., Park, Y.K., Jung, H.S., Kim, Y.B. & Kim, S.Y. (2010). Extract of Punica granatum Inhibits Skin Photoaging Induced by UVB Irradiation. International Journal of Dermatology,Vol.49, No.3, pp.276-282, ISSN 1365-4632

112. Patel AK, Michaud P, Singhania RR, Soccol CR, Pandey A. 2010. Polysacharides from probiotics: new developments as food additives. Food Technol Biotechnol 48:451–63

113. Perry P, Wolff S. New Giemsa method for the differential staining of sister chromatids. Nature 1974;251:156–8

114. Pirinccioglu, M.; Kizil, G.; Kizil, M.; Kanay, Z.; Ketani, (2012). A. The protective role of pomegranate juice against carbon tetrachloride-induced oxidative stress in rats. Toxicol. Ind. Health, DOI: 10.1177/0748233712464809,

115. Porrini, M.; Riso, P. Factors (2008). influencing the bioavailability of antioxidants in foods: A critical appraisal. Nutr. Metab. Cardiovasc Dis., 18, 647–650.

116. Poyrazoglu ,E., Gokmen, V. & Artik, N. (2002). Organic Acids and Phenolic Compounds in Pomegranates (Punica granatum L.) Grown in Turkey. Journal of Food Composition and Analysis,Vol.15, No.5, pp.567-575, ISSN 0889-1575

117. Poyrazoglu ,E., Gokmen, V. & Artik, N. (2002). Organic Acids and Phenolic Compounds in Pomegranates (Punica granatum L.) Grown in Turkey. Journal of Food Composition and Analysis,Vol.15, No.5, pp.567-575, ISSN 0889-1575

118. Proestos, C., Lytoudi, K., Mavromelanidou, O.K., Zoumpoulakis, P. andSinanoglou, V.J., 2013. Antioxidant capacity of selected plant extracts and their essential oils. Antioxidants 2: 11-22.

119. Pruthi, J.S. & Saxena, A.K. (1984). Studies on Anardana (dried pomegranate seeds). Journal of Food Science and Technology,Vol.21, No.5, pp.296, ISSN 0022-1155

120. Pundir CS, Rawal R. 2013. Determination of sulfite with emphasis on biosensing methods: a review. Anal Bioanal Chem 405:3049–62. Rasooli I. 2007. Food preservation—a biopreservative approach. Food 1:111–36.

121. Raffo, A., La Malfa, G., Fogliano, V., Madani, G.& Quaglia, G. (2006). Seasonal Variations in Antioxidant Components of Cherry Tomatoes (Lyco-persicon esculentum cv. Naomi F1). Journal of Food Composition and Analysis,Vol.19, No.1, pp.11-19, ISSN 0889-1575

122. Ramirez-Jimenez, A., E. Guerra-Hernandezand B. Garcia-Villanova. 2000. Browning indicators in bread. Journal of Agriculture and Food Chemistry, 48: 4176-4181.

123. Rada-Mendoza, M., A. Olano and M. Villamiel. 2002. Determination of hydroxymethylfurfural in commercial jams and in fruit-based infant foods. Food Chemistry, 79: 513-516.

124. Reddy MK, Gupta SK, Jacob MR, Khan SI, Ferreira. D (2007). Antioxidant, antimalarial and antimicrobial activities of tannin-rich fractions, ellagitannins and phenolic acids from *Punica granatum* L. Planta Med. May;73(5):461-7.

125. Rencuzogullari E,′ Ila HB, Kayraldiz A, Topaktas, M. 2001. Chromosome aberrations and sister chromatid exchanges in cultured human lymphocytes treated with sodium metabisulfite, a food preservative. Mutat Res490:107–12.

126. Ross SM (2009) Pomegranate: its role in cardiovascular health. Holist Nurs. Pract. 23:195-197.

127. Saad B, Bari MF, Saleh MI, Ahmad K, Talib MKM. 2005. Simultaneous determination of preservatives (benzoic acid, sorbic acid, methylparaben and propylparaben) in foodstuffs using high-performance liquid chromatography. J Chromatogr A 1073:393–7

128. Sadeghi, N., Jannat, B., Oveisi, M.R., Hajimahmoodi, M. & Photovat, M. (2009). Antioxidant Activity of Iranian Pomegranate (Punica granatum L.) Seed Extracts. Journal of Agriculture, Science and Technology,Vol.11, Supplementary Issue, pp.633-638, ISSN 1680-7073

129. Sampath, D. and Plunkett, W. (2001). Design of new anticancer therapies targeting cell cycle checkpoint pathways. Curr. Opin. Oncol., 13: 484-490.

130. Sarıkaya R, Solak K (2003) Benzoik Asit'in Drosophila melanogaster'de Somatik Mutasyon ve Rekombinasyon Testiile Genotoksisitesinin Aras ˌtırılması.GU" Gazi Eg˘itimFaku"ltesi Dergisi 23:19–32

131. Sarıkaya R, CˌakırS, (2005) Genotoxicity of four food preservatives and their combinations in theDrosophilawing spot test. Environ Toxicol Pharmacol 20:424–430

132. SAS, 2010. SAS/ STAT Users Guide for Personal Computers Release 9.1 SAS. Institute Inc. Cary and N.C,USA.

133. Sasaki YF, Kawaguchi S, Kamaya A, Ohshita M, Kabasawa K,Iwama K, Taniguchi K, Tsuda S (2002) The comet assaywith 8 mouse organs: results with 39 currently used foodadditives. Mutat Res 519:103–119

134. Schnuch A, Lessmann H, Geier J, Uter W. 2011. Contact allergy to preservatives. Analysis of IVDK data 1996–2009. Br J Dermatol 164:1316–25

135. Scognamiglio J, Jones L, Vitale D, Letizia CS, Api AM. 2012. Fragrance material review on benzyl alcohol. Food Chem Toxicol 50:5140–60

136. Sindelar JJ, Milkowski AL. 2012. Human safety controversies surrounding nitrate and nitrite in the diet. Nitric Oxide 26:259–66

137. Singh, D. & Singh, R.K. (2004). Processed Products of Pomegranate. Natural Product Radiance. Vol.3, No.2, pp.66-68, ISSN 0972-592X

138. Sebranek JG, Bacus JN. 2007. Meat products without direct addition of nitrate or nitrite: what are the issues? Meat Sci 77:136–47.

139. Seeram, N.P., Adams, L.S., Henning, S.M., Niu, Y., Zhang, Y., Nair, M.G. & Heber, D. (2005). In vitro Antiproliferative, Apoptotic and Antioxidant Activities of Punicalagin, Ellagic acid and a Total Pomegranate Tannin Extract are Enhanced in Combination with Other Polyphenols as Found in Pomegranate Juice. Journal of Nutritional Biochemistry, Vol.16, No.6, pp.360-367, ISSN 0955-2863

140. Seeram, N.P., Henning, S.M., Zhang, Y., Suchard, M., Li, Z. & Heber, D. (2006). Pomegranate Juice Ellagitannin Metabolites are Present in Human Plasma and Some Persist in Urine for up to 48 hours. Journal of Nutrition, Vol.136, No.10, pp.2481-2485, ISSN 0022-3166

141. Selvendiran, K., R. Padmavathi, V. Magesh and D. Sakthisekaran. 2005. Preliminary study on inhibition of genotoxicity by piperine in mice. Fitoterapia, 76(3-4): 296-300.

142. Shabtay A, Eitam H, Tadmor Y, Orlov A, Meir A, Weinberg P, Weinberg ZG, Chen Y, Brosh A, Izhaki I, Kerem Z (2008) Nutritive and antioxidative potential of fresh and stored pomegranate industrial byproduct as a novel beef cattle feed. J. Agric. Food Chem. 56:10063-10070

143. Sharma, A. and Sharma, A. (1980). Techniques. (3 th ed), Butterworth. London.

144. Suh H, Cho Y, Chung M, Kim B. 2007. Preliminary data on sulphite intake from the Korean diet. J Food Comp Anal 20:212–9.

145. Sudheesh . S and N. R. Vijayalakshmi, "Flavonoids from Punica granatum-potential antiperoxidative agents," Fitoterapia, vol. 76, no. 2, pp. 181–186, 2005. View at Publisher · View at Google Scholar · View at Scopus

146. Sumathy .R., Sankaranarayanan.S , Bama. P., Ramachandran .J..,M. Vijayalakshmi and M. Deecaraman (2013) Antioxidant and antihemolytic activity of flavanold extract from fruit peel of punica granatum. Asian J Pharm Clin Res, Vol 6, Suppl 2: 211-214

147. Sumner, M.D., Elliott-Eller, M., Weidner, G., Daubenmier, J.J., Chew, M.H., Marlin, R., Raisin, C.J. & Ornish, D. (2005). Effects of Pomegranate Juice Consumption on Myocardial Perfusion in Patientswith Coronary Heart Disease. The American Journal of Cardiology,Vol.96, No.6, pp. 810-814, ISSN 0002-9149

148. Stratil P, Klejdus B, Kub´ aˇ n V. 2006. Determination of total content of phenolic compounds and their antioxidant activity in vegetables–Evaluation of spectrophotometric methods. J Agric Food Chem 54:607–16

149. Tan, J. C. 1995. Sister Chromatid Differentiation and Exchaange in Chinese Hamster Ovary Cells in Culture. Pages 57-70, in Tested studies for laboratory teaching, Volume 16 (C. A. Goldman,Editor). Proceedings of the 16th Workshop/Conferencofthe Association for Biology Laboratory Education (ABLE), 273 pages.

150. Tavares RS, Martins FC, Oliveira PJ, Ramalho-Santos J, Peixoto FP. 2009. mParabens in male infertility—is there a mitochondrial connection? Reprod Toxicol 27:1–7.

151. Tiwari S. (2012). Punica granatum - A 'Swiss Army Knife' in the field of ethnomedicines. J Nat Prod ;5:4.

152. T¨ urko˘ glu S. 2008. Evaluation of genotoxic effects of sodium propionate and potassium propionate on the root meristem cells ofAllium cepa. Food Chem Toxicol 46:2035–41

153. Turk, G., Sonmez, M., Aydin, M., Yuce, A., Gur, S., Yuksel, M., Aksu, E.H. & Aksoy, H. (2008). Effects of Pomegranate Juice

Consumption on Sperm Quality, Spermatogenic Cell Density, Antioxidant Activity and Testosterone Level in Male Rats. Clinical Nutrition, Vol.27, No.2, pp.289-296, ISSN 0261-5614

154. Tsuyuki, H., Ito, S. & Nakatsukasa, Y.(1981). Lipids in Pomegranate Seeds. Nihon Daigaku No-Juigakubu Gakujutsu Kenkyu Hokoku,Vol.38, pp.141-148, ISSN 0016-5964

155. Yaseen, N. Y. ; Humadi, A.A.; Tawfiq, M.S. and Estivan, A.G. 1998. Cytogenetic studies on patients with chronic myclocytic leukemia. The Medical Journal of Tikrit Univ. 4 : 5-9 .

156. Yılmaz S, U¨ nal F, Aksoy H, Yu¨zbas,ıog˘lu D, C, elik M (2008) Cytogenetic effects of citric acid and benzoic acid on Allium chromosomes. Fresen Environ Bull 17:1029–1037

157. Yilmaz S," Unal F, Y" uzbas,io˘ glu D. 2009. The in vitro genotoxicity of benzoic acid in human peripheral blood lymphocytes. Cytotechnology 60:55–6

158. Yılmaz, Y., I. Çelik and F. Işık. 2007. Mineral composition and total phenolic content of pomegranate molasses. Journal of Food, Agriculture & Environment, Vol.5 (3&4):102- 104.

159. Yoshioka N, Ichihashi K. 2008. Determination of 40 synthetic food colors in drinks and candies by high-performance liquid chromatography using a short column with photodiode array detection. Talanta 74:1408–13

160. Yoshikawa K, Saito S, Sakuragawa A. 2011. Simultaneous analysis of acidulants and preservatives in food samples by using capillary zone electrophoresis with indirect UV detection. Food Chem 127:1385–90.

161. Yu¨zbas,ıog˘lu D, C , elik M, Yılmaz S, U¨nal F, Aksoy H (2006) Clastogenicity of the fungicide afugan in cultured human lymphocytes. Mutat Res 604:53–59

162. Valadares, M.C.; Perelra, E.R.T.; Benfica, P.L.; Paula, J.R. (2010). Assessment of mutagenic and antimutagenic effects of

Punica granatumin mice. Braz. J. Pharm. Sci., v.46, n.1, p.121-127,

163. Vally H, Misso NLA, Madan V. 2009. Clinical effects of sulphite additives. Clin Exp Allergy 39:1643–51

164. Vandevijvere S, Temme E, Andjelkovic M, Will MD, Goeyens L, Loco JV. 2010. Estimate of intake of sulphites in the Belgian adult population. Food Addit Contam A 27:1072–83

165. Vardin, H. & Abbasoglu, M. (2004). Nar Eksisive Narin Diger Degerlendirme Olanaklari. Geleneksel Gidalar Sempozyumu, Van, Turkiye, 23-24 September

166. Viuda-Martos, M., Fernández-López, J. & Pérez-Álvarez, J.A. (2010). Pomegranate and Many Functional Components as Related to Human Health: A Review. Comprehensive Reviews in Food Science and Food Safety,Vol.9, No.6, pp.635-654, ISSN 1541-4337

167. Wang, R., Ding, Y., Liu, R., Xiang, L. & Du, L. (2010). Pomegranate: Constituents, Bioactivities and Pharmacokinetics. In: Fruit, Vegetable and Cereal Science and Biotechnology,da Silva, J.A.T. (Ed), pp. 77-87,ISSN 1752-3419, Global Science Books

168. Wang L, Zhang X, Wang Y, Wang W. 2006. Simultaneous determination of preservatives in soft drinks, yogurts and sauces by a novel solid-phase extraction element and thermal desorption-gas chromatography. Anal ChimActa 577:62–7.

169. West T. Atzeva M. and Holtzman D. 2007. Pomegranate polyphenols and resveratrol protect the neonatal brain against hypoxic-ischemic injury. Dev. Neurosci., 29 : 363-367.

170. WHO, World Health Organization. 2000. Concise International Chemical Assessment Document. Benzoic acid and sodium benzoate. Available from: http://www.who.int/ipcs/publications/cicad/cicad26_rev_1.pdf.Accessed2013 July.

171. Ulca P, Atamer B, Keskin M, Senyuva HZ. 2013. Sorbate and benzoate in Turkish retail foodstuffs. Food Addit Contam B 6:209–13.

172. Ulca P," Ozt" urk Y, Senyuva HZ. 2011. Survey of sulphites in wine and various Turkish food and food products intended for export, 2007-2010. Food Addit Contam B 4:226–30

173. Zarei, M., Azizi, M. & Bashir-Sadr, Z. (2011). Evaluation of Physicochemical Characteristics of Pomegranate (Punica granatum L.) Fruit during Ripening. Fruits,Vol.66, No.2, pp.121-129, ISSN 0248-129

174. Zeiger E, Anderson B, Haworth S, Lawlor T, Mortelmans K(1988)Salmonellamutagenicity tests: IV. Results from the testing of 300 chemicals. Environ Mol Mutagen 11:1–158

175. Zhang, L., Fu, Q. & Zhang, Y. (2011). Composition of Antociyanins in Pomegranate Flowers and their Antioxidant Activity. Food Chemistry,Vol.127, No.3, pp.1444-1449, ISSN 0308-8146

176. Zeiger E, Anderson B, Haworth S, Lawlor T, Mortelmans K (1988)Salmonella mutagenicity tests: IV. Results from the testing of 300 chemicals. Environ Mol Mutagen 11:1–158

177. Zengin N, Y" uzbas‚io˘glu D, " Unal F, Aksoy H. 2011. The evaluation of the genotoxicity of two food preservatives: sodium benzoate and potassium benzoate. Food Chem Toxicol 49:763–9.

I want morebooks!

Buy your books fast and straightforward online - at one of the world's fastest growing online book stores! Environmentally sound due to Print-on-Demand technologies.

Buy your books online at
www.get-morebooks.com

Kaufen Sie Ihre Bücher schnell und unkompliziert online – auf einer der am schnellsten wachsenden Buchhandelsplattformen weltweit!
Dank Print-On-Demand umwelt- und ressourcenschonend produziert.

Bücher schneller online kaufen
www.morebooks.de

OmniScriptum Marketing DEU GmbH
Heinrich-Böcking-Str. 6-8
D - 66121 Saarbrücken
Telefax: +49 681 93 81 567-9

info@omniscriptum.com
www.omniscriptum.com